Oak and Ash and Thorn

'A joy of a book and a delight to read.'   *The Great Outdoors*

'A tender hymn to the trees, a manifesto for a woodland society, a contemporary gazette of ideas and attitudes radiating into the future like annual rings from the original pith… In this lyrical, informative, unashamedly arboreal propaganda, one man's walk in the woods can inspire a generation.'

Paul Evans, author of *Field Notes from the Edge*

'Fiennes is the best of guides, gently, eloquently and with a fierce humour telling a sad story – relating chapters of fascinating detail to brighten his tale and quoting the poets as he goes.'

John Wright, author of *A Natural History of the Hedgerow*

'Passionate and thoughtful in exactly the way the best nature writing should be… the woodlands of Britain have found their perfect advocate.'

Hugh Thomson, author of *The Green Road into the Trees*

'A wonderful wander into the woods that explores our deep-rooted connections – cultural, historical and personal – with the trees.'

Rob Cowen, author of *Common Ground*

'Peter Fiennes really can see the wood for the trees – he blends mythology, natural history and a sense of righteous anger to produce a paean of praise to our ancient woodlands and modern forests, and the life support system they provide.'

Stephen Moss, author of *Wild Kingdom: Bringing Back Britain's Wildlife*

'A wonderful and enlightening meander through the different issues surrounding our woods and trees… One moves from the dark and the unsettling into sunlit glades, frequently stopping to connect with moments of beauty but also the shocking truth of the threats facing our woods today. Conversational in tone, it is shot through with humour which often had me laughing out loud.'

Rebecca Speight, CEO of The Woodland Trust

About the Author

PETER FIENNES is the author of *To War with God*, a moving account of his grandfather's service as a chaplain in the First World War. As publisher for *Time Out*, he published their city guides, as well as books about London's trees and Britain's countryside.

Oak
and Ash
and Thorn

The Ancient
Woods and New
Forests of Britain

Peter Fiennes

ONEWORLD

A Oneworld Book

First published by Oneworld Publications, 2017
This paperback edition published 2018
Reprinted, 2018

ISBN 978-1-78607-321-1
eISBN 978-1-78607-167-5

Illustration on p. vi, *Under the Yew Tree*, © Flora McLachlan

Typeset by Tetragon, London
Printed and bound in Great Britain by Clays Ltd, Elcograf S.p.A.

Oneworld Publications
10 Bloomsbury Street
London WC1B 3SR
England

MIX
Paper from
responsible sources
FSC® C018072

Dedicated with love to my woodland walking companions,
Anna, Natalie, Alex, Esme, Biddy and Bonnie
Some more willing than others

Contents

'Here, in fact, is nothing at all
Except a silent place that once rang loud,
And trees and us – imperfect friends, we men
And trees since time began; and nevertheless
Between us still we breed a mystery.'

EDWARD THOMAS, *FROM* 'THE CHALK-PIT'

Introduction

'If you go down to the woods today you're sure of a big surprise.'

JIMMY KENNEDY, 1932

Like most people in Britain, I no longer have any daily connection with its woods. There are a lucky few who live and work among the trees, but mostly we go about our lives as though they are not there – sealing ourselves in at work and at home, hurrying from car park to shop, keeping the wild at bay. This is a change from even fifty years ago, when the boundaries between our towns and the woods were more ragged and blurred – and certainly from one hundred years earlier, when most of us would have known someone who worked or foraged in the woods, or we would have done so ourselves.

Today, we have less woodland cover than almost any other European country, despite four decades of hard planting by the Woodland Trust and other dedicated charities. For a couple of generations, we even seemed determined to get rid of the woods entirely, ripping up and poisoning the native broadleaf trees and replacing them with close-packed rows of fast-growing conifers. And yet, despite all this, I don't think we were ever going to let the woods go – and nor were they ready to leave us. They still fill our childhood stories, call to us from the edges of the cities, and permeate our dreams. Like millions of others, I was horrified when the government announced in 2010 that it was selling off our public woods – even though at the time, in all honesty, I had no idea that there even was such a thing as a public reserve of forests. Later,

I found myself wondering: what kind of demented ideologue would want to sell them off?

This thwarted government sell-off must have preyed on my mind, because a few years later I found myself gripped with a desire to find out what was going on in the woods. Perhaps it was because I'd lived in London for so long, and was adrift from the woods of my childhood, but I felt I needed to find out what shape the woods were in – and who was looking after them. From what I read, they always seemed to be *under threat* – from roads, high-speed railways, housing, neglect, golf courses, ignorance, greed and the convulsing climate. So that's how I ended up spending a year in the woods. Not literally, I'm afraid – there's nothing here on how to survive on a diet of squirrels and tubers – but for twelve months I visited as many woods as I could, from small copses to the new national forests, and I steeped myself in the poetry, science, folklore, history and magic of woodlands.

The original plan was to split the book into chapters that would follow the themes that interested me: conservation, ownership, conifers, history, magic and myth, our fear of the woods, climate change, woodland legends, childhood, and our current obsession with what is 'native' or 'alien'. It seemed that there was also something to say about Britishness (or Englishness) and belonging. Along the way I wanted to learn about every species of British tree (and hunt down some ancient ones) and even find a scrap of the original wildwood, untouched by the busy workings of humanity. But of course the woods and the trees are not so easily contained. The lines between the chapters became confused, the roots entangled, and I did what any child could have told me I shouldn't have done: *I left the path*. Still, I think the book is better for it – and the original pattern can be easily traced, like a medieval woodbank showing through the scrub of a regenerating forest.

My year in the woods covered the seasons and the book is roughly chronological, beginning and ending in the same wood in Herefordshire, close to the English/Welsh border. I was surprised by what I found – the damage being done, but also the many inspiring people working hard to nurture, preserve, spread and replenish our woods. Above all, I was left with a new sense of urgency. Time is running out and we desperately need to enthuse new generations to love and protect the woods, although that is only going to happen if we throw them open to as many people as possible. Despite everything, I am very hopeful. There are millions of us who care deeply for woods – and even if sometimes it might seem as though we are beset by bigger, more devastating problems, just try to imagine a land without trees.

Kipling was right:

> Surely we sing no little thing,
> In Oak, and Ash, and Thorn!

I

The woodland thicket overtops me,
the blackbird sings me a lay, praise I will not conceal;
above my lined little booklet
the trilling of birds sings to me.
The clear cuckoo sings to me, lovely discourse,
in its grey cloak from the crest of the bushes;
truly – may the Lord protect me! –
Well do I write under the forest wood.

ANGLO-SAXON POEM

Timber!

Croft Ambrey, late May

It is early summer in the woods above Croft Castle in Herefordshire. The silver birch leaves are turning a deeper green, their bark a smooth, tender shade of white. Snug among the leaves, the pale green catkins are furled and ripening like a million caterpillars trembling on the brink of release. There is only a narrow strip of birch here, by the edge of the path, a sprinkling of young trees standing at a crossroads in the heart of a gloomy expanse of conifer plantations, but enough light has reached the woodland floor at this point to mean that there is also grass, bracken, red campion and even a dash of bluebells. Butterflies waltz in the sunlight. And beyond the birch, the conifer plantations are spread far across the hillside, the forest floor dark in the midday sun, dead brown needles lying thick on the dismal ground.

I am gazing at a newly erected National Trust sign:

VISITORS TO CROFT CASTLE AND ITS SURROUNDING WOODLAND MAY BE SURPRISED TO SEE A NUMBER OF TREES BEING FELLED OVER THE NEXT SIX MONTHS, BUT THIS IS A MAJOR STEP TOWARDS REINSTATING THE BEAUTY SPOT'S HISTORIC WOOD PASTURE.

THE FORESTRY COMMISSION IS REMOVING 70 ACRES OF NON-NATIVE CONIFERS FROM THE CENTRAL PART OF CROFT WOOD AS PART OF ITS PLANNED WOODLAND MANAGEMENT.

CONIFER PLANTATIONS CONFLICT WITH HOW THE LANDSCAPE LOOKED UP UNTIL THE MID-20TH CENTURY.

Over the decades, many of us have learned to hate conifer plantations. Even the Forestry Commission now seems to regard them with a glum and sheepish dissatisfaction, despite the fact that it was they who were responsible for most of the planting in the first place. Certainly, in the decades after 1919, when the commission was formed as a response to a wartime shortage of pit props and trench cladding, it was unstinting in its efforts to secure the national supply of timber. 'Non-native' conifers were selected – they are fast-growing and regular in their habits – and were spread with aggressive abandon across the country. Neglected farmland was chosen first, followed by remote expanses of peat and the thin-soiled uplands, before the commission finally turned its Sauron-like gaze on Britain's last isolated remnants of broadleaved woodland. Nothing was safe (there was almost no legal protection), as ancient woods across the land were grubbed out, drenched in chemicals, uprooted and replaced by orderly rows of Sitka and Norway spruce, Japanese larch and Corsican pine.

We know all this now – although anyone who was paying attention also knew it at the time. R.S. Thomas, the Welsh poet who preached and brooded just across the border from Croft Ambrey, had his own bleak perspective of the plantations:

> I see the cheap times
> Against which they grow:
> Thin houses for dupes,
> Pages of pale trash,
> A world that has gone sour
> With spruce.

FROM 'AFFORESTATION'

'A world that has gone sour with spruce' pretty much sums it up. Mind you, Thomas's solution was to return the land to men and their sapling-stripping, monocultural sheep, an idea that would give most modern-day environmentalists the vapours. The poem was written in the 1960s – and the destruction and regimented planting had become especially frenzied in the years after the Second World War. Oliver Rackham, our pre-eminent woodland writer, dubbed the years 1950–75 'The Locust Years' – although, to be more accurate, what these particular locusts devoured they also replaced: thousands of hectares of diverse ancient woodland gave way to monotonous lines of spruce; teeming, self-regenerating, wildlife-rich woods became closed, sterile circuits of industrialized timber production.

In those years, even people who loved their local woods may have felt that their destruction was somehow necessary. Perhaps a leftover spirit of sacrifice from the war, along with a new obsession with the statistics of economic growth, convinced enough people that scientifically managed plantations were the right, futuristic solution to Britain's perceived timber shortage. More likely, most people didn't care enough and just sort of assumed that someone else knew better. It's true that immediately after the Second World War there was less woodland in Britain than at any time since the last Ice Age, but there had been many similar panics about a national shortage of timber over the centuries. Unfortunately, this was the first moment that the panic coincided with the tools (and the imported seeds and trees) to transform our landscape within a generation. No one stopped to consider whether we actually needed to be self-sufficient in timber (it was centuries since we last had been), they just demanded more trees. Lots and lots of them. And, whereas in the past, the oak tree had been the patriotic tree of choice for the would-be forester, this time it was the spruce.

If you want a glimpse of the national mood during these disastrous years, then a good place to start is *Trees, Woods and Man* by

H.L. Edlin, first published in 1956 as part of Collins's iconic 'New Naturalist' series. Edlin loved his trees, and lamented the dying out of the traditional woodland crafts (and even the disappearance of our native woodland), but what he really wanted to see was the efficient 're-establishment of forests by modern methods'. One of his most frequently used woodland words is 'crop', because in his eyes trees were really no different from wheat, beans or potatoes. Trees were timber. And a wood was nothing more than a production line of trees. Anything else that grew or lived in the wood was getting in the way of the harvest.

'The general practice of the British forester is to clear fell his woodland and to start his new crop from scratch,' he notes approvingly, but 'How closely does the forester plant his young crop, and what type of trees does he use? On the average he sets the trees in rows, five feet apart each way, using about 1,750 to each acre (or about 1,500 after allowing for roads, rides, and like gaps).'

A sense of urgency pervades the pages; there is no time to wait for nature to run its course, or for our slow-moving, clearly rather hopeless native woodlands to get their act together. There is pressing, grown-up business at hand: 'Vast woods, ripe for slow regeneration in this way [i.e. self-seeding], have had to be slaughtered within a few months,' he tells us, and as for Britain's fragile and irreplaceable reserves of peat: 'The Forestry Commission, following the lead of a few pioneering private landowners [oh yes], has steadily been developing ways and means of draining peat bogs, and finding trees, such as the spruce, that tolerate them.' In his vision, learned in his years as a rubber planter in Malaya, every scrap of what was deemed to be non-productive land had to be put to good use. It's not a crazy idea – no doubt Edlin and others had witnessed enough shortages, hardship and starvation to make them impatient of any obstacles to economic growth or bleatings about aesthetics – but it was symptomatic of the forces that led to the almost total annihilation of Britain's native woodland. There has been a bitter and hard-fought backlash, and a change in policy at the Forestry Commission, but we remain stuck with a veneration of productivity and growth, along with a corresponding confusion about anything that cannot be measured as useful. We may have left the Locust Years but, if anything, the state of mind that gave rise to them is even more entrenched. Why else would anyone even have to explain what a wood is *for*?

Many (English) nature writers and landscape historians, surveying the destruction, seem keen to blame the Germans for all this. At some point in the eighteenth century the Prussians got it into their heads that the forests could be regulated, the trees paraded into neat lines and the productivity of woods maximized on strictly scientific, Enlightenment principles. It was all a question of measurement. Oak trees and most other broadleaves (messy, erratic, slow, inclined to host wildlife and far too tolerant – if not downright neighbourly – towards other species of tree) were destroyed and replaced by the conifer plantations that we now think of as a quintessentially German landscape. Within a couple of generations, the famous German oak forests had gone. Edlin (with his suspiciously Teutonic name) was just following the mood of the times. It wasn't that he was unaware of the beauty of trees or the glory of the woods (he wrote with elegiac intensity about them, even as they disappeared from view), but he would hardly have expected such intangible ideas to take precedence over the nation's economic needs. In the 1965–67 'stocktaking' at the start of *Trees, Woods and Man* (a census of woodlands carried out by the Forestry Commission) he proudly displays a table showing the state of Britain's woodlands, alongside the same 'stocktaking' from 1947. It shows that in the period between 1947 and 1967 (the years when the locusts really got going) the amount of woodland in Britain had actually increased by twenty per cent, but that the increase was entirely down to the mass planting of conifers.

It was the woodlands that the Forestry Commission deemed 'unproductive' – broadleaf coppices and native woods – that were neglected or destroyed and in just twenty years had fallen (from an already shockingly low base) by another thirty per cent. For centuries, the coppices, the broadleaf woods that were managed on a cycle to produce most of our fuel and much of our timber, had been our most 'productive' of all; now they were dismissed as a waste of space: as too diverse, slow and complicated. The frenzy of destruction continued deep into the 1970s and continues – openly, insidiously – to this day. The idea that woodland – all of nature – is something to be accounted for, with stocktakings and harvests, and what's more that it is incapable of looking after itself and is something that needs to be *managed*, has become deep-rooted and instinctive. There have always been other voices. John Fowles raged against the 'disastrously arrogant male dominated religions, which supposed man to be in God's image and duly appointed him, like some hopelessly venal and ultimately crazed gamekeeper, the steward of all creation'. But most of the time nature lovers and naturalists were dismissed as an irritating irrelevance and an encumbrance to progress and profit. The pendulum may be swinging, but it would be naive to assume that it is heading in the right direction.

The path up from the crossroads among the conifers leads towards Croft Ambrey, the remnants of an Iron Age fort. This land – these woods – has been worked by people for over 2,500 years and, as naturalists always seem quick to tell us, there is not a scrap of unblemished woodland left on these islands: it has all at some point been chopped down, managed, replanted or in some way pawed over by someone. I refuse to believe that – it's a grim, desiccated thought, rather like knowing that every inch of Britain (the world!) can be conjured up on a screen, courtesy of Google Earth. Who would want that? So one thing I have in mind is to try to find a little parcel of woodland where it might at least be possible to imagine a world before people got so indefatigably busy with their axes, fires and crops. Or at least one where only the most light-footed people have ever trod, stepping with care over the fallen trees.

That place is not here. Many of the conifers, I now notice, have been marked with red crosses, a sign that their days in the forest are numbered. It's a pleasing thought, but also unexpectedly and absurdly troubling, triggering a whisper of unease about the rounding up of 'non-natives' and the cleansing of unwelcome and invasive aliens. People do get very agitated about introduced species: rhododendrons, Corsican pines, parakeets, grey squirrels, Japanese knotweed, Chinese mitten crabs, horse chestnuts, sweet chestnuts, Spanish worms, Spanish bluebells, Dutch elm bark beetles, mink, muntjacs and rabbits. Foreigners every one. Brought here by the Romans, borne back triumphantly by Victorian plant collectors, escaping from farms and gardens to rampage through our land, overwhelming or just simply eating the insipid and degenerate locals. Some of them are undeniably successful, these imported plants, fungi and animals.

The native woodland that the National Trust is hoping to cherish on this hillside will include any species that once managed to scramble over the land bridge that still connected Britain to mainland Europe about seven thousand years ago, just before the ice melted and the waters rose, cutting us off from France and creating an island. There is still some debate about which species of tree qualify, but there are probably twenty-six of them. The last over the causeway may well have been the box tree, which has always struck me as a rather unlikely native, more suited to a sun-dazzled Mediterranean mountainside than the backdrop for a tense picnic in Jane Austen's *Emma*. One tree that didn't make it was the sycamore (although given the enthusiasm with which it has spread ever since it was introduced you'd have to think this was an oversight – perhaps it woke up late one morning somewhere near Dieppe to find the drawbridge raised and all its so-called friends already colonizing the White Cliffs of Dover). It has made up for lost time since – a fact that seems to enrage some people. I think we should take our bearings from Barbara Briggs, author of the 1934 children's book *Our Friendly Trees*:

> Sycamores are such common trees all over England that it is difficult to believe that they are really foreigners and were not brought here from Europe till the fifteenth century. They seem to like this country very much, for they have settled down as if they had been here always, in town and suburban gardens, parks and fields, and even on the salty sea-coast where no other large-leaved trees can live… We are very often inclined to look down on common trees and flowers, and to think that others are more beautiful just because they are rare. But if you watch a sycamore all the year round, the unfolding leaves, the blooming flowers, the ripening fruits, and the shadow patterns of its foliage on the sunlit grass, you will realize that it is a very handsome tree, besides being a homely and friendly one.

There are, predictably, a few sycamore trees at the edge of the conifer plantations, although there is no word from the Forestry Commission about what they have in store for this stubbornly prolific tree: 'First they come for the conifers…' In fact, there are also plenty of self-seeded conifers here, on the dark edge of their own plantation, jostling for the sunlight. This surprises me, even though it shouldn't, but I hadn't stopped to think that conifers would spread all by themselves, without being planted into neat rows by a friendly forester. There is, in reality, no turning back. What we've introduced will persevere and prosper if it can and – across the planet, and in all sorts of unexpected ways – we are reaping what we have sowed. The fertility of trees *is* surprising, though. Ever since the early twentieth century, the oak tree, so generously fecund with its acorns, has stopped being able to propagate *inside* woodland. The acorns catch and root and grow on the edges of woods, or in the open – and in wood pasture, indeed, unless they are bothered by sheep – but they will not take hold within a wood of any density. Oliver Rackham has a tentative explanation, blaming the accidental introduction of an American fungus, but it's troubling to think that our most iconic tree is growing old childless. And perhaps Rackham shied away from a more disturbing possibility, one that he was too scrupulously scientific to voice, that our woods and trees are somehow knocked off balance and wounded more deeply than we know.

There's a big old oak tree near the top of Croft Ambrey, about twenty paces from the rough path. It's just about the last tree you see as you leave the woods and reach the cleared ground near the defensive earthworks at the summit. It's possible that the people who lived in this place from about 2,500 years ago may have kept a few trees scattered around their hill fort, whether as living objects of veneration or perhaps because they ringed the walls with planted hawthorn, ready to be cut down as an extra layer of thorny defence. Trees grow more expansively and die younger when they're not crowded in a wood, so it's hard to tell how old this particular oak tree might be (and I'd have to cut it down and count its rings to have any chance of getting close to guessing), but it could be over 400 years old. Its great-grandfather may well have been a sapling in the last days of the Iron Age people who lived in this fort and then gave it up to the invading Romans. Trees' lives are lived to a different rhythm and timescale to ours, which is one reason we love them. They help us look up and out.

I say that the oak tree is about twenty paces from the path, but that's just its trunk. One immense branch starts low at the base, bends close to the earth, then forges up and over where I'm standing. It could be a medium-sized tree in its own right – and the strength that must be needed to keep it horizontal is incalculable. A blackbird sits singing at its tip, far from the centre; I wonder if it has any idea that it is sitting in a single tree, and not a copse. This is probably not the kind of thing blackbirds 'think' about. An oak tree this size is its own planet, giving succour to over 300 other species – everything from bacteria to buzzards. Nettles, brambles and goose grass (or 'sticky willy' as I once used to call it, before people looked at me strangely), even a small hawthorn, all flourish in its shade. The trunk is split some way from the ground, but still bullishly alive. As I gaze at its rough surface I am shaken by how huge this tree has become. Not just its thick grooved trunk, but look up and there are dozens of branches that twist and swirl skywards, and even bigger branches, which I hadn't seen from the path, flung far out up the hill. Its roots are braced on the slope, planted deep like medieval bulwarks. I could worship this tree. We should all worship this tree. And just as I am thinking that, I hear the words very clearly in my head (I know it sounds ridiculous, but I do) 'noli me tangere'. *Don't touch!* Why did I think that? Was it from the tree? Do trees talk? Well, yes, it's likely that they do (if not in Latin, let's hope), but more of that later.

Many people were upset when the Conservative Party adopted the oak tree as its new party logo in 2006, and then painted it red, white and blue. Or even blue and green (this was before they dispensed with all the 'green crap' and apparently gave up trying to affect any of the life-threatening dangers that are poised to engulf every living thing on the planet). It wasn't just because it was the Conservatives doing this: it doesn't seem right that any political party, of all things, should be claiming a tree (*the* national tree) as their own. But of course we can see what they were trying to do. John Ruskin wrote that 'the man who could remain a radical in a wood country is a disgrace to his species' – and I think that's right. Why would any of us, standing in wood country, want to change a thing? Maybe remove the conifers (planted by radicals of all parties, no doubt), perhaps pick a posy of wild flowers or chew on a blackberry, but that's about it. The Conservatives, in claiming the oak as their own, are hoping to show that they are standing at one with the unchanging certainties of the past – a past that connects seamlessly with a bright, burnished and utterly secure future – and the woods, and the countryside, and indeed the planet itself, have nothing to fear from any slight changes they may be about to make.

In reality, the Conservatives are just part of a continuum of ecologically destructive governments, who generally have had other things to think about than the health of our forests. And okay, yes, there is something 'conservative' about the oak tree. When young King Arthur ('the Wart', as he is known) in T.H. White's *The Sword in the Stone* flies up into the forest (he has been transformed into an owl, obviously), he finds the oak tree has this to say to him: 'I am a conservative, I am; and out of my apples they make ink, whose words may live as long as me, even as me, the oak.' Oak trees are big and loveable and slow-moving and *daunting* – and they live for ever and we want to hug and worship them. People may think of

oak trees as 'conservative', but that's only because, in a supreme
act of anthropomorphism, it is hard to imagine an oak tree doing
anything skittish, or getting involved in a get-rich-quick scheme, or
selling off the National Health Service to its friends in the private
health sector. They are 'conservative' with a small 'c' – they don't
like change. And if you're confused by T.H. White's reference to oak
'apples', they are those hard, light-brown, marble-sized balls you
see on oak trees, not the seed but the home of the gall wasp larva:
yet another species that makes its home on the oak. The black fluid
distilled from the gall was used for centuries as ink for writers' quills.

Oak trees also have the most delicate, sinuous green leaves
in spring, so slight and vulnerable when set against the massive
weight of their elephant-grey wood. It seems like a good day, as
the earth bustles into early summer, to wonder which tree has
the most beautiful spring leaves. There are hornbeams here, far
from their Home Counties heartland. The sunlight drifting down
through their knitted branches gives the leaves a lovely fresh green
sheen. The lime tree's leaves are even greener – and so soft, it's like
they're made of the faintest crepe paper. Samuel Taylor Coleridge
once, waving his wife and friends off on a country walk, was forced
by a bad burn on his foot to spend the day in a 'lime-tree bower'.
No doubt in considerable pain, presumably self-medicating with
enormous draughts from a flask of opiates, it was only because he
couldn't move that he had to soak himself in the loveliness close at
hand and so was able to 'keep the heart awake to Love and Beauty!':

> Or in this bower,
> This little lime-tree bower, have I not mark'd
> Much that has sooth'd me. Pale beneath the blaze
> Hung the transparent foliage; and I watch'd
> Some broad and sunny leaf, and lov'd to see

The shadow of the leaf and stem above
Dappling its sunshine!

COLERIDGE, *FROM*
'THIS LIME-TREE BOWER MY PRISON'

What he then goes on to say is that beauty is everywhere, and that he didn't need to go roaming the cliff tops with his friends to find it. He might also have reflected that it was his constant philosophizing, late-night revelations and incessant yakking that drove his long-suffering wife to throw scalding milk on his restlessly pacing feet, thereby forcing him to stay behind brooding under the lime tree. Although, if she hadn't lashed out in a rage, we would never have been blessed with one of the English language's most beautiful poems.

The hazel, one of my favourite trees, is rather dowdy at this time of year. It's better to see them earlier, when the yellow-dusted catkins emerge on a late-winter's day, to the utter relief of anyone who has been yearning for a sign that winter surely cannot last for ever. There's a big ash tree at one end of Croft Ambrey and its leaves are struggling through rather later than the other trees' foliage, but now they're here they've come in sprays and spumes, a frothing surge thrust up from the small, uptipped black buds (Coleridge's spirit is lurking…). The ash can't compare to the big sticky buds of the chestnut, nor the pink and placidly erotic sycamore, but it is undeniably generous with its fishbone leaves and profusion of small purple flowers.

It is, of course, impossible to choose a favourite springtime leaf. There are a couple of maple trees here, the native field variety, and their newborn leaves are hanging limp and clean like so much spar-kling laundry. It is a fresh, invigorating sight, but it's still probably better to catch the maples during their autumn display, even if they are not exactly Canadian in their splendour. No doubt there are some beech trees lurking, and the birch is sweet and dainty earlier in the year, but I'm distracted by the sight of a dozen hawthorn trees in full bloom. The hawthorn looks so dead through the winter, its grey bark sunk into dreary gloom, but when the leaves do finally emerge (so much later than expected) they are flat and glossy and pulsing with optimism. And then the flowers come out – and keep coming. So many of them, drenching every branch and twig, the whole tree so bushily alive. Is the hawthorn just a little too busy? Yes, yes, we get it: leaves, flowers, perfume in the air, insects homing in – well done. But isn't it all just a little needy? I cross the hawthorn from my 'Best Spring Leaf' list and decide that the task is absurd. And who would even want to choose their favourite tree?

The views from the top of Croft Ambrey are extraordinary. They say you can see six counties from here (on a clear day is the predictable qualifier – not an especially likely occurrence this close to Wales). There's a quarry on the north side, half a hill scoured and gouged grey, bringing stone for houses, gardens and roads; beyond it the green fields and hedgerows head for the Shropshire Hills. In the other direction – towards the Malvern Hills or, with half a twist, the Wye Valley and Brecon Beacons – there are bare hills with square patches of woodland, all of it, as far as I can see, conifer plantations. I really don't want to find that I am accompanied on my every walk by this conifer obsession, but I am not ready or able to leave it alone. The New York poet Eva Salzman, who found herself living in Kent in the 1970s (from Tribeca to Tunbridge Wells!), had this to say:

> I follow an ordnance map and find
> frightening rows of straight and vacant pines.
> The earth as barren as the rugs
> people in my nearby town put down. Medicine

sting of pine. Listen there, hear nothing. No bird sings,
I'm told that insects are the only living things
in that Forest Commission flat. And slugs.

FROM 'ENDING UP IN KENT'

The poets, in other words, have been complaining for a while, but it
has been a long time since enough people cared what a poet thought
about anything important. Did we ever? More than two hundred
years ago, William Wordsworth was grumbling about the planting
of pines in the Lake District and flung his hat in a rage at a larch
tree. He would be distraught to see the place now. We are not oak
trees and we don't notice the full sweep and scope of the changes
we have unleashed. We can only comprehend the immediate detail:
the loss of a wood here, a river dammed and poisoned there, a
water meadow drained, England's last wolf cornered and killed.
If we could live as an oak for eight hundred years, and watch the
landscape as the busy mass of humanity scurried and scraped at
its surface, the villages emptying, coalfields rising and falling, the
last pockets of wilderness tidied away, we would feel the agony of
our loss. The conifers are just a distraction.

Although... as I walk by a different route down from the summit
of Croft Ambrey, I come to a place where the Forestry Commission
has already started clearing its conifers. It's a huge area of violent
devastation. Almost all the trees are gone and a churn of macerated
larch covers the ground for hundreds of metres in every direction.
The fury of destruction has flung branches and trunks among the
still-living conifers further down the hill. Here, all the trees are
marked with green or orange paint, awaiting their catalogued fate.
Dead wood is everywhere. Bark. Branches. Chips and shreds. Roots
ripped from the ground. Did the conifers at least fulfil their original destiny and get taken to be made into pine tables? It's hard to

imagine that there was anything left to take, given the vast quantity of biomass that remains, stripped and sprayed across the hillside. There are signs telling us to stick to the path, but I scramble over the massed debris to take a look at one single pine that has been left, a tall tree with no branches (it had once struggled up to the light in fierce competition) that stands amidst the desolation like a raised middle finger. Up yours and fuck you, it seems to be saying. I never asked for this.

The new plantings don't seem to be doing very well. There are small, square, fenced areas where the new broadleaf saplings are being prepared for their life as 'wood pasture' – an area of woodland with roughly twenty per cent tree cover, given over to animal grazing. Many of the trees will be pollarded – cut on a seven- to ten-year cycle at a spot on the trunk just out of the reach of hungry cattle and deer – and once the new trees have grown this should be a beautiful place. There are young hornbeam, holly, maple, some thorn trees and, I'm sure, oak. But right now many of the new saplings seem to be struggling ('struggling' as in 'dead' in most cases). I wonder if the soil has been irredeemably sucked dry or turned to acid by the conifers, and I hope it's unlikely. The Forestry Commission will just have to try again next year. One broadleaf tree has got a head start. It must have been growing in the conifer plantation for the last thirty years, hidden in the forest, a thin, spindly, apologetic kind of thing – possibly a wild service tree – but now it has been set free to grow and spread. Under a fork of its bottom branches is a tiny, self-seeded fir tree, missed by the men with the machines. I feel pleased about that, although it's not very tidy. But perhaps our obsession with tidiness is something we need to address. We have been managing our forests in the same way we have been trying to manage our economies and our lives. Keeping everything in rows. Clinging to the grid. Harvesting the frail, the superfluous and the irregular. Chopping it all down and starting again. As the Forestry Commission concedes defeat and embraces the wild, perhaps in the wider world the men with clipboards will learn that not everything can be reduced to simple economic rows. Perhaps they will stop measuring success by how much they have pruned and saved, and how many people they have transplanted or trimmed, and instead start thinking about creating an environment where everyone can flourish. Or perhaps they won't…

A pair of buzzards drifts and circles in the cool spring air above the lacerated pines. In the distance I see there are also three young-ish oak trees, sheared at the top and leaning against one another. They are set solid in a deranged explosion of bluebells and that, of course, is why we want to be rid of the conifers: we want our bluebells back, not to mention the ferns, the cowslips, the anemones, primroses and wild garlic. Someone has said that individually the bluebell is not an especially interesting flower, but that's not true. It has a soft and rather brave charm, although – and of course – en masse they glow and shine with a radiant blue light. Is this the return of 'Old England', with only the native trees and plants untouched in the slaughter? A beating back of the Locust Years? We have to hope so, although there are still plenty of conifers left, even here, and it's fanciful to imagine we can ever return truly to what we once had. But the idea of this imagined future wood pasture lifts the heart. I can even gaze at the conifer plantation down the hill without irritation. It looks almost Alpine here, I realize, and an Alpine landscape is one of the joys of this world. But just not here. If I didn't know that these conifers were alien and planted as a short-term crop, would I care so much? Show most people these woods and I'm sure they'd prefer the bluebell-laden broadleaves; but they wouldn't necessarily be feeling the visceral cultural antipathy to the conifers that I've managed to generate.

We are all infused with the views of our times, and a hundred years ago even the most English of writers would have found much to love in the pine forest:

> A wind sways the pines,
> And below
> Not a breath of wild air;
> Still as the mosses that glow
> On the flooring and over the lines
> Of the roots here and there.
> The pine-tree drops its dead;
> They are quiet, as under the sea.
> Overhead, overhead
> Rushes life in a race,
> As the clouds the clouds chase;
> And we go,
> And we drop like the fruits of the tree,
> Even we,
> Even so.

GEORGE MEREDITH, 'DIRGE IN WOODS'

The pine woods here are planted more thinly than most, the trees taller. It would be wrong to say that the birdsong ceases, but it is now replaced by the noises of the sea, the soothing sound of the wind in the top branches. Perhaps the conifers are nearing their time for harvest, but they have been thinned and this means that there are at least brambles and ferns on the forest floor. If these trees were planted on what was once an ancient wood, then when they go there's a chance that the woodland flowers and fungi may return more easily. But to bring back an ancient wood's true diversity takes centuries (so far as we know, but who has had the time to check?), which is why only woodland that predates the year 1600 has been classified as 'ancient'. Frankly, that's not ancient at all (ask an oak), but it's a start. And it means that the loss of even the smallest scrap of ancient woodland should be counted as a national tragedy. An international tragedy, given that without our trees we won't be able to breathe. Or eat. Trees clean our land and our air, they scrub our rivers and our soil, they keep the oceans alive, they soak up our effluence, toxins, greenhouse gases and poisons. Without trees we will all die. Britain has less tree cover than almost any European country – we are contending with Ireland and The Netherlands at the bottom of a dismal league. We need to hoard and cherish every wood we still have – and we need to plant more. Many many more. Woods are not just for walking in. They are our lungs, our refuge, our playgrounds, our solace. They are the generous providers of fuel, timber, food, energy and life. And for us, they are a 'kingdom free from time and sky', says Louis MacNeice in 'Woods'. They let us dream of other ways of living.

I don't see why our whole country can't be covered in woods. And all the buildings, warehouses and homes, the factories, roads and railways, they should all be topped by mist-laden, life-giving trees as far as the eye could see. There would be huge oak forests surrounding our towns and beech trees standing thick and tall on the high streets. There would be orchards in the wastelands and vast fountains of ivy would foam from the walls of every public building. We need the trees more than we know, and we need them everywhere, from the most remote valleys to the heart of the teeming cities.

2

'*Les forêts précèdent les peuples, et les déserts les suivent.*'
(Forests precede people, and the deserts follow them.)

CHATEAUBRIAND

City Limits

London, in fact, is a forest. All it takes to get the official classification is twenty per cent tree cover, and London easily passes that mark, especially if we include the verdant outer fringes of Epping and Richmond. Maybe we should be saying that every city in the UK is officially 'wood pasture'. Some – Bristol and Sheffield spring to mind – do feel as though they retain a connection with the outlying countryside, even if Sheffield Council is in the news for butchering its famous street trees. You will also need to ignore the cars, roads, exhaust fumes, pavements, shops, lighting, litter, noise and – a killer point – the invisible toxins. Not that there aren't enough of those running into our fields and rivers. But take a look around your nearest city. There are trees everywhere. Not only where you might expect them, in the squares, parks, gardens and commons, but also on almost every street and pavement. In London, ornamental almonds and thick ropes of wisteria adorn the roads and smother the pastel houses of Primrose Hill; and on the Old Kent Road, at the very bottom of the Monopoly board, there are dozens of trees – blasted by pollution, sideswiped by buses, but standing strong and tall.

Of course, almost none of these London trees have arrived by chance, so although Londoners live in a forest, it's not a very *wild* forest and perhaps it would be more accurate to say that we live in an empty plantation, from which the undergrowth has been scrubbed away. There is much more to a forest than just a bunch of trees. We must miss the primordial woods, though, because we

have been busily encouraging it back ever since we built and then
tore down the city walls.

I'm a Londoner, although I didn't grow up here. Not many of
us did. There's an age-old idea that there's a fixed and impermea-
ble border between 'the city' and 'the country', dating back to the
Greeks, no doubt, and reinforced by the European city states, but
it has always seemed like an absurd fantasy. The country – shall
we call it 'nature'? – permeates every inch of the city, no matter
how much pollution or bleach we throw in its way. Likewise, but
with more noise and fuss, the city has spread outwards in the past
two hundred years, the suburbs and streets of London absorbing
hundreds of one-time villages, farms, fields and woods. We all
know what someone means when they say that they live in 'a city',
'a village', 'the suburbs' or 'the country', but the boundaries are
porous and the underlying assumption is dangerous. We can't live
our lives imagining that we are not part of nature. It's not something 'out there' that needs to be managed and tamed – or more
likely forgotten about, just so long as there is food in the shops and
no flooding in the streets. Our cities are not separate from nature,
and nor are we. We just like to pretend otherwise, although who
wouldn't want to think that hunger, disease, wolves – even death –
could all be stopped at the city gates?

I live close to Clapham Common in south London. It's a broad,
flat expanse of green, a typical city park, lined with avenues of
mature horse chestnut, lime and London plane trees, and popular
with Sunday footballers, dog walkers, joggers and, in the tiny scrap
of woodland on the west side, the occasional cottaging MP. It really
is a very small wood, with a busy main road on one of its edges (the
road is close enough to *every* edge, in all honesty), but it still manages
to exhale beauty and wonder in every season. I have walked here
in the dead of winter – fallen leaves dark and slippery underfoot,

ice cracking on the rutted ground – and felt a long way from the shrieking city. And I've been here on that very first unexpected day of spring, when winter is still growling over its meal, but all of a sudden there's a scent in the air, a change of rhythm, a faint pulse closer to the surface of the trees, and you know that everything is going to change, now, and fast.

In fact, looking up on this winter's day, there's a sparkle of black-thorn blossom against the clear blue sky. There's wild cherry, too, and the ivy gripping the trees has a lighter sheen, even if it hasn't yet started this year's climb to the top. Shoots of phosphorescent grass are making tentative appearances at the base of a horse chestnut, whose young buds are gleaming as though they've just been licked. On the other side of the sodden path, in the stacked brambles and clumps of nettle, there's no sign of new life: all I can see are last year's leaves, still limply hanging on. We've just been through the warmest winter 'since records began', which sounds like a good thing, but it's not. It's one in a series of increasingly urgent warnings that we are choosing to ignore. I miss the frosts and the snow, the crystal clarity of winter, the snap of ice, and these few damp months of soggy chill are a miserable excuse for a season. Perhaps appropriately, a finch's eager little song is drowned out by the latest passing plane, flying low and heavy over the wood.

It is still winter, despite the premonitions of change. Most of the trees are in lockdown, saving energy. The wood's solitary beech tree is hunched and hibernating, as is a moribund ash, its keys hanging in thick dusty bunches from every branch; en masse they look like small grey pineapples, although they scatter and crumble when touched. There's also a large London plane tree here, quite rightly dominating this small London wood. I've grown to love these trees. They're a hybrid of the American and the oriental plane, but were first bred in Spain, so go by the name *Platanus hispanica* – a real mixture and viewed with suspicion by the native woodland purists. They won't propagate in London without intervention, poor things, despite the eye-catchingly hairy balls that hang in impressive quantities from their winter branches. They fall in the spring, once the new leaves start to appear, only to be gobbled up by another successful (and clearly very fertile) alien, the grey squirrel. The first London plane is said to have arrived in England in 1663 when it was planted in the University of Oxford Botanic Garden, but we've been spreading the tree frenetically ever since, especially in London, once we discovered its prodigious resistance to pollution. Where other more sensitive trees are overwhelmed by smoke, smog and industrial waste, the London plane just shrugs off another layer of bark and carries on soaking up the miasmic gases through its huge leaves.

The plane trees of Berkeley Square, from which, once upon a time, nightingales sang, are said to be the oldest trees in central London. (There are some truly venerable oaks further from the centre.) The planes have been here about 250 years and over the centuries have evolved into weird, fantastical shapes, absurdly tall, but lumpen and distended at their sprawling bases. The dirty grey, scaled trunks are constantly sloughing off extra layers, which is probably not the best Mayfair look, although there's a chance it may come into fashion in a post-apocalyptic world. And that's something that will be upon us sooner than we would like, given the London and national governments' strenuous efforts to avoid having to clean up the city's air. Over forty thousand people in the UK are choking and dying unnecessarily every year because of poor air quality. Imagine if that figure were deaths from measles, or bombs. Perhaps our only option, given that we've been abandoned to our fate by authorities too scared to challenge the status quo, is to plant more London plane trees. A vast number of them. We owe these cooling, soothing, quietly heroic giants a great debt.

The London plane in the little patch of woodland on Clapham Common is probably about a hundred years old, but they are fast growers and there is already a prodigious amount of timber in its thick trunk and erratic branches. In winter you can more clearly see the zigzag growth of its twigs: after each leaf bud is formed the new shoot heads off in the opposite direction. The nearby notice from the 'Friends of Clapham Common' rather sniffily notes that its wood has 'no known economic use', which has been the kiss of death for most species that have bumped up against humanity over the centuries, but the plane has shown itself to be useful. You can see the anti-pollution measures in action on this and every other London plane: sheets of greenish bark are peeling from the trunk, revealing a dun yellow undercoat. It's a friendly tree, late to leaf in the spring and tolerant of others. There's a small hawthorn seeming to grow out of this plane's roots and snuggling up against the older tree's trunk. They could almost be one tree.

The Clapham Common copse (as no one else calls it) is best in the summer, when the leaves muffle the roar of the planes and hide the view of the cars and trucks. There are two old lime trees at the centre, such comforting trees at this time of year, with their tender leaves, soft low branches and welcoming, lichen-coated, gently grooved trunks. One of them must have been pollarded a long time ago, or perhaps it's a natural sundering, but about twelve feet from the ground the main trunk twists and splits into three, and then the three new trunks soar dizzyingly skywards. I go to find these trees one midsummer's day, close to dusk, walking fast through the surrounding tangle of hazel scrub, and there, spread out on the early summer pasture, at ease in his very own woodland glade, is an almost naked, thin white man in tight, brilliant-white underpants. The air is heady with acacia blossom. I don't ask him his political party – I suppose it's possible he's not an MP – but he looks absolutely right here, shifting drowsily in the gloaming on his bank of grass. Woodland glades are magical places and things can happen in woods that turn the world upside down. Oberon knew this:

> I know a bank whereon the wild thyme blows,
> Where oxlips and the nodding violet grows
> Quite over-canopied with luscious woodbine,
> With sweet musk-roses, and with eglantine:
> There sleeps Titania some time of the night,
> Lull'd in these flowers with dances and delight.

> SHAKESPEARE, FROM ACT II, SCENE I,
> *A MIDSUMMER NIGHT'S DREAM*

Some years before these lines were written, Henry VIII led his pregnant wife, Catherine of Aragon, out of London and up to a glade in the woods below Shooters Hill. There he set up camp and spent the day lording it over a bunch of actors playing 'Robin Hood', 'Friar Tuck' and 'Maid Marian', while hundreds of extras milled around, dressed in Lincoln green, carrying longbows and wondering at what point they should discreetly retire so that their king could grab his wife, or even 'Marian', and start rattling the bushes. Or maybe he was mooning over Catherine and she slapped his pale little hands away and told him to wait until they were safely delivered of their child. It would be no surprise to learn he comfort-consumed a hell of a lot of venison that day.

Even if Henry wasn't getting any, woodlands have always been the scene of love-making, especially in the days when privacy was hard to come by. The great festival was May Day, and the highlight was the raising of the maypole, when all the young women and men would head to the woods at dawn to gather the May flower (bushels of frothing hawthorn blossom) to adorn and then dance around the welcoming pole. You don't need to be a dyed-in-the-wool Freudian to see where this is leading, and the early Puritan Philip Stubbes raged against such lubricious behaviour in his *Anatomie of Abuses* (1583):

> All the young men and maids, old men and wives, run gadding over night to the woods, groves, hills and mountains, where they spend all night in pleasant pastimes... the chiefest jewel they bring from thence is their May-pole, which they bring home with great veneration... And this being reared up, with handkerchiefs and flags hovering on the top, they strew the ground round about, bind green boughs about it, set up summer halls, bowers and arbours hard by it. And then they fall to dance about it... I have heard it credibly reported (and that *viva voce*) by men of great gravity and reputation, that of forty, threescore, or a hundred maids going to the wood over night, there have scarcely the third part of them returned undefiled.

It sounds like a lay-by on the A6 outside Preston. I especially enjoy that '*viva voce*', the huddled whispering together of 'men of great gravity and reputation' as they contemplate the 'rearing up' of the maypole, its binding with 'green boughs', and all those pert young things dancing and grabbing each other in the woods. It's enough to get an honest Puritan quite worked up.

The practice of maypole dancing has largely died out, as people have moved to the cities and taken up clubbing, although you'll still see poles in some villages and primary schools at the right time of year. In Thomas Hardy's *The Woodlanders*, written three hundred years after Philip Stubbes's fulminations, the young girls of Little

Hintock village head to the woods on Midsummer's Eve, followed not-so-subtly by most of the local men. When the clock of Great Hintock strikes midnight the girls cast a handful of hemp seeds to the ground, mutter an incantation and hope to be confronted by a vision of their future husband. What they get instead, as they try to sprint home, is a wholly unsurprising encounter with the men, and there is much 'giggling and writhing' and 'desperate struggle'. The heroine, Grace Melbury, has been educated out of her natural connection to the woods, and instead of being waylaid by the earthy but maddeningly shy Giles Winterborne, who lives at one with the forest, is instead gripped by the effete outsider Dr Fitzpiers, who claims her as his own with a kiss. Once that's done, and she's heading for home with 'the moon whitening her hot blush away', Fitzpiers bounds after the very willing 'hoydenish maiden of the hamlet', Suke Damson (indeed), and they have it away in a hayrick.

You can judge the worth of the characters of *The Woodlanders* by their feelings for the woods. On the one hand there's the bibbling idiot Fitzpiers, and the bored and overwrought Mrs Charmond, who in the damning words of one old village crone is 'the wrong sort of woman for Hintock – hardly knowing a beech from a woak [*sic*]'. On the other hand, there's Giles and his wild kindred spirit (but for some reason unfanciable), Marty (who ends up having to sell her lustrous hair to the dreadful Mrs Charmond). The beautiful Grace Melbury, meanwhile, moves tragically between the two camps, yearning to travel to civilized foreign lands with Mrs Charmond, but beset by visions of Giles who rises 'upon her memory as the fruit-god and the wood-god in alternation: sometimes leafy and smeared with green lichen, as she had seen him amongst the sappy boughs of the plantations: sometimes cider-stained and starred with apple-pips, as she had met him on his return from cider-making in Blackmoor Vale, with his vats and presses beside him.' Grace never does get the chance to rub down Giles's lichen-smeared flanks, with his vats and presses beside him, and instead marries Fitzpiers, making both of them (in fact, everyone) miserable. Just to spell it out, 'the casual glimpses which the ordinary population bestowed upon that wondrous world of sap and leaves called the Hintock woods had been with these two, Giles and Marty, a clear gaze'. There is truth in the woods, but only some can see it. At least, according to Hardy.

Hardy wrote about woods with longing and love, and although he grew up in Dorset, and returned there once he was making a living from his writing, he also lived in London for some years, and for three of them (1878–81) in his late thirties he was renting a home on Trinity Road, one of the main routes out of the south of the city. *The Return of the Native*, his first significantly successful novel, was published while he was living there. It's strange to think of Hardy, such a deeply rural writer, living in London, but of course Trinity Road was not the traffic-scarred highway it is today. And Hardy is surprising. He wrote longingly of being:

> Heart-halt and spirit-lame,
> City-opprest,
> Unto this wood I came
> As to a nest

FROM 'IN A WOOD'

But he then finishes the poem by telling us that he'd rather be living amongst people: at least they smile, and talk and can even, just occasionally, be loyal and friendly to one another. Trees, he believed, fight to suppress their rivals (and even their own kind) in the struggle for light, nutrients and forest supremacy. Honeysuckle chokes saplings, holly bushes recoil from thorn trees, sycamore and oak shoulder each other aside, and ivy hangs like a noose in the woods. There's a strip of woodland on the edge of Trinity Road, not far from Hardy's old home, and there's at least one tree living here that must have been around in Hardy's day. One oak in particular stands in a clearing just yards from the road, its trunk dark with urban grime (and probably harbouring only about a dozen of the possible three hundred-plus species normally associated with the oak), but there's no doubt that Hardy himself would have brushed past it on his way to the shops. He obviously felt sorry for it.

> Here you stay
> Night and day,
> Never, never going away!
>
> Do you ache
> When we take
> Holiday for our health's sake?
>
> Wish for feet
> When the heat
> Scalds you in the brick-built street,
>
> That you might
> Climb the height
> Where your ancestry saw light,
>
> Find a brook
> In some nook
> There to purge your swarthy look?

HARDY, *FROM* 'TO A TREE IN LONDON'

Hardy seems to imagine that the tree will have to leave London in order to reconnect with its ancestors and 'purge' its 'swarthy look', but there were trees in London long before humans and their houses – just as once upon a time there were lions, hippopotamuses, hyenas and super-elephants, basking and bickering on the banks of the River Thames. The climate has changed many times since then, as has the fauna, but I suppose it is *possible* that nothing has grown here, in this small wood close to Hardy's old home, other than native British trees and plants, ever since the most recent Ice Age retreated northwards.

The ghosts of an older, pre-industrial world are everywhere in London. You can stick a pin into a London street map and the odds are that you will hit one of them. I try this on my local map and strike 'Meadow Road', an obvious rural ghost. In the year 1800 the Earl of Dorset had an estate here and was making yet more money by infilling eastwards from the South Lambeth Road. So Dorset Road leads into Meadow Road and Meadow Mews drifts off into a postwar estate. It sounds idyllic. There's no doubt that Londoners like to be reminded of meadows and woodland and perhaps that's why so many of our streets are named after woods and trees and orchards and vales: they're a memento of what was once there – a stand of trees, a pasture, a patch of rough ground, a working copse. Sometimes, of course, they're just names dreamed up by the developer, or they remind the builder of what he left behind in his Dorset home. A bucolic name certainly won't hurt the house prices. But I'm afraid for once the name jars, standing in 'Meadow Road' by a derelict pub, with only a few scrappy street saplings for company – although someone has taken the trouble to plant some flowers at the base of the trees. It's a hopeful sight.

If you scratch the surface of any London road, and take a look at who has lived there over the years, you will always find deep-rooted connections to the natural world, and not just in the street names. In Meadow Road, for example, opposite the Ashmole Estate at the north end of the street, is a lovely set of early Victorian houses and cottages, the first buildings ever to settle upon the Earl of Dorset's water meadow. Their fronts are painted perky shades of pastel and there's enough wisteria to suffocate an elephant. In the 1940s the extra-wide house at number 8 was lived in by John Menlove Edwards, pioneering Welsh mountaineer, restless and off-beam psychiatrist and would-be conscientious objector (his application for CO status was refused by a Liverpool board in 1940). Edwards was one of the pre-eminent rock climbers of his age. He liked to climb in wet, difficult conditions – a damp Welsh scree slope over an Alpine rock face – with no pitons or ropes. He loved overhangs where the hand-holds were treacherous and crumbling, and was the first to scramble up many now-famous Snowdonia climbs. As his friend and fellow climber Geoffrey Winthrop Young put it, he was 'serpentine and powerful as an anaconda coiling up loose or wet overhangs'. Edwards was gay (seriously, 'Menlove'? – who says our names don't govern our destinies?), at a time when homosexuality was illegal, and this may well have contributed to the desperately sad decision to take his own life with a cyanide pill in 1958. Meadow Road must have seemed a long way from those Welsh mountains.

The natural world is all around us in the cities. But so, inescapably, are traces of war. There's an ancient hawthorn tree at the front of Branksome House, the four-storey housing block that marks the edge of a large 1950s estate built on bombed-out land. Its well-tended lawns ('no dogs'), playground ('no ball games') and window boxes are in much better repair than the shabby late-Victorian houses opposite. One of these, number 74, a nondescript, grey, three-storey terraced house, was home to James Foster Barnsley and his wife, Susannah Maria, from about October 1915 – although

James, on his wedding day and in a flurry of excitement, managed to give as his address number 47. We know this because James Barnsley was a private in the 2nd Battalion of the Essex Regiment and was killed in action aged twenty-seven on 1 May 1917 – and the local Stockwell War Memorial has been assiduously researching all its dead. James's brother, William Charles, was also killed in the war, although seven of their siblings survived.

They would almost certainly have known the Bunn family at number 35. John Bunn, three years older than James Barnsley, enlisted with the King's Royal Rifle Corps and died of his wounds on 10 August 1917. John's 'C' Company, 10th Battalion, was thrown into the first days of the Ypres offensive and, at some point in that gruesome bloodbath, possibly in a failed attempt to cross the River Steenbeck in full view of German machine-gunners, John floundered and died. Nature and war. Their ghosts slip and glide through the fabric of London. And so it seems right to ask: does a love of the beauties of nature make war and violence (or the pull of war, the allure of violence) any less likely? To put it another way: if we are not exposed to nature, if we are starved of woods and mountains, clean rivers and fresh horizons, are we more likely to turn on each other and our surroundings? The creators of London's parks certainly thought so, and most people would agree, including the princes and despots of Europe in the eighteenth and nineteenth centuries, who looked enviously at Britain's relative freedom from rioting workers and commissioned *Englischer Gartens* across the Continent in an attempt to placate their own restless proletariats. They believed (and so do I!) that there's something intrinsically good about a wood, a park, a stretch of grass. God knows, they're better than stinking Dickensian terraces or the modern, antiseptic, high-rise glass and steel fantasies. We are all Romantics now. But I think of Hitler (sorry to bring him in), sitting and smiling on the

terrace of his gorgeous Alpine retreat; of Göring striding through
the ancient Polish forest, brushing past the sleepy oaks, loosing off
armaments at the last of the European bison; of Henry VIII, even,
reclining in some rose-covered arbour, humming 'Greensleeves' and
plotting another intimate murder. And what about Pol Pot in the
jungle, Charles Manson in his commune, the cabin in the woods
with the madman sharpening his axe? It doesn't add up. Despite
this, I am certain that the more time we spend in close contact with
nature – in the woods! – the more our humanity will grow. It's just
hard to be sure about that. At least George Orwell, in *Some Thoughts
on the Common Toad*, thinks it might be true:

> I think that by retaining one's childhood love of such
> things as trees, fishes, butterflies and – to return to my first
> instance – toads, one makes a peaceful and decent future
> a little more probable, and that by preaching the doctrine
> that nothing is to be admired except steel and concrete,
> one merely makes it a little surer that human beings will
> have no outlet for their surplus energy except in hatred
> and leader worship.

A short mortar-shot from Meadow Road is Lansdowne Gardens,
the first London home of Edward Thomas, the writer and poet
of birdsong, woodland and meadows, who was killed in the Battle
of Arras in early 1917. He was born here on 3 March 1878,
joining his family in a cramped rented flat, before moving south
into the new-build streets of Clapham and some slightly more
spacious accommodation. At some point in his childhood, probably during frequent family trips to Wales and Wiltshire, Thomas
imbibed his lifelong love of nature, and by the time he left Oxford
University he was determined to dedicate his life to writing about

it. He even published his first book, *The Woodland Life*, while still an undergraduate, and for the next twenty years he kept up a ferocious rate of production, churning out hundreds of literary reviews, essays, travel pieces and books. He only came to poetry late, urged on by his friend, the American poet Robert Frost, and much of it was written after he had joined the army, and was living in the village of Steep in Hampshire. It was Frost's insight that Thomas's prose – and, above all, the jottings in his notebooks – would be brought most vividly to life if it were distilled into poetry.

I head for Steep, by car, on a dirt-drab January day. If it had been May, I might have empathized more with Thomas's opening line in his essay 'Spring on the Pilgrims' Way' that 'Even the motor road is pleasant now when the nightingales sing out of the bluebell thickets under oak and sweet chestnut and hornbeam and hazel' – but that seems unlikely in these road-ruined days. As I drive, I find myself marking off street names and road signs that are named after trees or woodland. There are dozens of them (although, I'd rather not be too specific about my route, because I don't want to be told by some London know-it-all that if only I'd taken the second left at Lytton Grove I'd have shaved twenty minutes off my journey time). There couldn't be a clearer illustration of how much has changed in the hundred years since Thomas came this way. I pass a series of ghostly woods (Knightwood Crescent, Southwood Drive, Woodside Close, Largewood Avenue, Woodland Way, Collingwood Avenue, Birchwood Lane, Firdene, Westcoombe); memories of trees (Elmbridge Avenue, Broomhill Road, Oakdene Drive, Oakleigh Avenue, Beech Close, Elm Lane, Beechcroft Drive, Hazel Grove, Hazel Bank, Broad Oaks); and countless 'Groves', 'Meadows' and 'Copses'. There's even a 'Wilderness Road'. All of it now paved.

So this is what happened
to the names of the trees!
I heard them fly up,
whistling, out of the woods.
But I did not know
where they had gone.

WENDELL BERRY,
'WALNUT ST., OAK ST., SYCAMORE ST., ETC.'

'London' does not want to end. The houses and the streets, drives, closes, avenues, shops and malls, the buildings and the roads, coagulate at the side of the route. It does end, of course: we have our green belt and the manicured, fiercely protected fields of Surrey (which is also, rather surprisingly, the UK's most wooded county); but this experience of change – much of it, at the time, unwelcome – has been the background hum to all our lives for generations. Tobias Smollett raged about 'the great Wen', the spreading cancerous lump of London, in 1771; and in 1852 Wilkie Collins had this to say about an 'unfinished' part of London (that is now probably one of its most lusted-after – and central – neighbourhoods):

We reached a suburb of new houses, intermingled with wretched patches of waste land, half built over. Unfinished streets, unfinished crescents, unfinished squares, unfinished shops, unfinished gardens, surrounded us. At last they stopped at a new square, and rang the bell at one of the newest of the new houses. The door was opened, and she and her companion disappeared. The house was partly detached. It bore no number; but was distinguished as North Villa. The square – unfinished like everything else in the neighbourhood – was called Hollyoake Square.

I noticed nothing else about the place at that time. Its
newness and desolateness of appearance revolted me,
just then...

COLLINS, FROM *BASIL* (1852)

Edward Thomas, in his more gentle fashion, remarks only on
'the enormous, astonishing, perhaps excessive, growth of towns,
from which the only immediate relief is the pure air and sun of
the country'. It is hard to fight a feeling of gloom on this choked
road south. Of course, this is a hellishly difficult problem. We
urgently need enough houses – and no one likes a NIMBY ('not
in my back yard'). But that is also a duplicitous term, invented
by a PR company working for the noxious owners of engorged
corporations, living lives of untrammelled ease many miles from
the land they hope to parcel up and profit from. Easy, green land,
with no complications. Just a few trees and fields. No, I don't think
most people do want a supermarket with parking for a thousand
at the edge of their village, no matter how many crèches and clock
towers it provides. Nor do they want more new housing than can
be comfortably absorbed by their roads, buses and schools, just
as they don't need some multinational energy company fracking
the hell out of their fields. That makes most of us YIMBYs –
yes, I would like some control over what happens to my own
back yard.

But I don't want to think about that now. Not yet. I want to get
to Edward Thomas's woods.

'Northfield' is the name of the wood opposite the little church
at Steep – the 'field' in the name an indication, but not conclusive
proof, that at some point in the past this was pasture or arable

land. In January, a waterlogged path wends its way through the trees towards 'The Hangers', the high, chalk, wooded ridges that must at one time have inspired someone, in a paroxysm of inspiration, to name their village 'Steep'. We've had a lot of rain and the way is narrow and circumscribed by threadbare winter hedges, the ground in between sticky with grey mud. I'm moving forwards with a series of hops and skiddy lunges, treading when I can on the banks of drab russet leaves that have drifted up against the edges. The woods seem to be energetically man- aged, with recent hazel coppicing and clearances, and timber stacked in beetle-friendly piles. Sky, trees and mud; it's a grey landscape, with just a couple of evergreen yew trees showing their muted winter coats. A robin puffs itself up on a gatepost, a smudge of colour. But that's about it until I notice the holly trees. There are dozens of them growing low in the understory, more bushes than trees, their leaves a sharp varnished green, the massed berries pulsing a deep blood-red. They're not tall: holly can grow to ninety-foot high and in some parts of the country it can overwhelm other species, but here it is growing low in the shadow of oaks, ash and hazel.

Midwinter is the holly's time, when the leaves of other trees fall and it seems to emerge from the woods. It is, as it says in *Gawain and the Green Knight*, 'greenest when groves are gaunt and bare'. At the winter solstice, holly is the 'holy' tree, the Holly King, ruling the sleeping world, only giving up the crown reluctantly to the Oak King, who reigns supreme through the summer. Or so I'm told. And I also know that if you throw holly leaves (or maybe the branches) at a wild beast it will kneel at your feet. Pliny the Elder tells us that. And you should cut your holly staff or wand with care and only after offering the appropriate thanks and libations: the holly is sacred and does not like to be taken for granted. But if you get it right, the holly provides a powerful witch's wand, potent with spiky male energy. In fact, men looking to attract a female partner should carry a few holly leaves with them. Holly, as incense or a tincture, can re-energize the stalest marriage bed. It's also worth strewing a few holly leaves under your pillow at night if you want to get a glimpse of your future, but do not do this lightly. Holly can lead you to the underworld. Nor should you, on any account, leave any holly inside the home after Twelfth Night: you will attract evil spirits. In Ireland it is bad luck to plant a holly tree too close to your home; in England the opposite is true – it will protect you from lightning and malicious faeries. In either country it is bad luck to chop down the tree. Instead, drink a cup of holly tea and you will find your jealousy and agitation subsiding. Do not, though, drink or consume the berries, they are poisonous to children and deeply upsetting to adults, even though John Evelyn, the magisterial seventeenth-century author of *Sylva: or A Discourse of Forest-Trees and the Propagation of Timber in His Majesties Dominions*, suggests swallowing 'a dozen of the mature berries... to purge phlegm without danger'. I think we are on safer ground following his advice on how to plant a holly hedge. Evelyn is also extremely and unsettlingly detailed about how to make 'Birdlime' out of the bark of the holly, a clagging substance that was smeared on the branches of trees in order to trap songbirds. One final word of advice: if you have just got married, then bring some holly leaves over the threshold of your new home. Men should bring the spiky leaves, women the smooth ones. Whoever does that will rule the roost for the rest of the marriage (although if you both bring leaves, it's not entirely clear what will happen – perhaps you'll be divorced within a week).

Holly, then, is king of the winter woods. Its top leaves are generally without spikes (brides-to-be have some climbing to do), but the lower leaves have evolved to grow strong and spiny in order to repel cattle. (But not deer, which munch through the things enthusiastically and must have tongues like hobnailed boots; perhaps the fact that they can do this with impunity should make us doubt the theory – or perhaps what has happened is that the deer's tongues have evolved faster than the holly's leaves.) Its berries are really fruit (with four stones), and they're digested and spread far and wide by the hungry winter birds. It's not really the only sign of green life in a British wood at Christmastime (there's juniper, yew, box, Scots pine and ivy), but the holly bears the crown:

> The holly and the ivy
> Now both are full well grown,
> Of all the trees that are in the wood,
> The holly bears the crown.
> O the rising of the sun,
> The running of the deer,
> The playing of the merry organ,
> Sweet singing in the quire.

The wood of the holly grows slowly and produces a heavy, white timber, which is often used for chess pieces or the handles of the whips of coachmen. Like the wood of the box tree, holly sinks in water. Indeed, if you are travelling after nightfall, always take your holly-handled whip with you to ward off evil spirits. According to H.L. Edlin in *British Woodland Trees* (1944), 'holly is of no importance as a timber tree, but is useful for hedges and ornamental planting. It will not thrive in smoky towns, where all evergreens tend to become "nevergreens".' This may be his only recorded joke, but the holly is a somewhat hysterical tree. Perhaps someone should make Mr Edlin a nice mug of hot holly tea.

At the foot of the Ashford Hanger, there are four massive stools of once-coppiced ash, now left to run high and free. These ash trees would have been cut every seven to ten years, but the regular harvesting of their timber for fence and hop poles, staves and posts, must have ceased decades ago. People who know and love woods will often point out that we need more coppicing – and the sound of a chainsaw in a wood is a happy sound. It is undeniable that the woodland landscape we cherish, especially in the south of England – the open glades with their primroses, foxgloves and bluebells – can only flourish if the tree cover and undergrowth are regularly cleared and the canopy is opened to bring sunlight to the forest floor. The axeman brings life to the woods.

These ash trees have been left to make their own way and they are already colossal. A coppiced tree lives far longer than one that has been left unpruned (sometimes, we're told, 'for ever'), although one of these ash trees is now threatened by a mid-sized oak growing out of the middle of its base. One day the oak will probably split and shatter its host, but for now they are living cheek-by-jowl, giving me a few moments of confusion as I try to untangle what kind of tree I am staring at. There's a small heart-shaped hole in one of the trunks, perfect, I hope, for nesting birds or bats. Behind the ash there is more holly. They are bigger, more dominant trees in this part of the wood, sharing space with a number of thick-scaled, prehistoric-looking Scots pine. A wood pigeon is sitting high in one of them, and it takes off, its wings clattering like a briskly shaken bundle of sticks.

The walk to the top of Ashford Hanger is, indeed, steep. Near the top there's a memorial to Edward Thomas, a slab of stone placed here in tribute by Walter de la Mare some years after Thomas was killed (by a bullet or a shell) while lighting his pipe in a hole in Arras. The plaque on the front reads:

THIS HILLSIDE

IS DEDICATED TO THE MEMORY OF

EDWARD THOMAS

POET

BORN IN LAMBETH 3RD MARCH 1878

KILLED IN THE BATTLE OF ARRAS 9TH APRIL 1917

AND I ROSE UP AND KNEW

THAT I WAS TIRED

AND CONTINUED MY JOURNEY

How nice, how apt, that a hillside is dedicated to Thomas. Is he the only poet to have such an honour? As far as I know, the rest of them are lying in obscure churchyards, or mouldering in Westminster Abbey, but Thomas lives on in the yew and the oak trees, the soft grasses and the chalky scree of his favourite hill. He believed that certain places were cherished and protected by their own 'little gods of the earth', and that 'it is a loss of a name and not of a belief that forbids us to say to-day that sprites flutter and tempt there among the new leaves of the beeches in the late May light'. He is right. We no longer have a name for the inexplicable magic of places, at least not one that we can utter without apology. But I think we know that there is such a thing as the spirit of a place, its *genius loci*, and I can tell you that this particular hillside is alive with the spirit of Edward Thomas.

If you're still with me, then you know 'Adlestrop':

> Yes, I remember Adlestrop –
> The name, because one afternoon
> Of heat the express-train drew up there
> Unwontedly. It was late June.
> The steam hissed. Someone cleared his throat.

No one left and no one came
On the bare platform. What I saw
Was Adlestrop – only the name
And willows, willow-herb, and grass,
And meadowsweet, and haycocks dry,
No whit less still and lonely fair
Than the high cloudlets in the sky.
And for that minute a blackbird sang
Close by, and round him, mistier,
Farther and farther, all the birds
Of Oxfordshire and Gloucestershire.

THOMAS, 'ADLESTROP'

Thomas conjures magic. It's a simple list – meadowsweet, willow-herb and grass, steam hissing from the express train – but he brings us with him to that station platform in late June. We can smell the engine and hear the birds. We can see the misty hills of England stretching far into the distance. We don't just want to be there. We *are* there. And he's clever because he gives us something we want to believe, that we live in a land of boundless beauty, where over the next hill there are more hills, more woods and meadows, high clouds and birdsong. It never ends. It fills me with longing to find the woods of my imaginings, a fairy-tale forest that goes on for ever, its trees and groves surging to the horizon, further than anyone can travel, smoke drifting up from the clearings, wolves and bears roaming safely in the wilderness.

Anyway. Enough soldiers travelled to the front carrying that crazy vision of England in their heads and hearts, and like Edward Thomas many of them never came back. I sit on a bench to the side of his memorial stone, facing down the hill, and try to imagine what he would have made of his view today. The yew trees are still here, growing thick and green on either side; and in the middle-distance there are small, leafless woods, overgrown hedgerows, mud-green fields and a few rooftops, shining silver in the January drizzle. There's one house with a rather grand garden, its large lawn tightly trimmed and punctured by some ornamental fir trees; but apart from that there are very few homes to be seen, let alone villages or even towns. There is, though, a bottle-green caravan in one of the fields, framed by some gratifyingly misty hills.

The view may be more or less the same, but when I think to listen for it, I realize that there is almost no birdsong. Or rather, it's here (and, of course, it's midwinter so I'm not expecting much), but what there is has been nullified by the sounds of the traffic on the nearby A3. This noise, unencumbered by summer's leaves, amplified by the valley, is a constant background moaning, so persistent that it's almost unnoticeable. But stop and listen. It's incessant. The rolling thunder of the cars, the scream of the revving motorbikes, the hellish payload of lorries howling along the unseen road. It is relentless. Not only that, what this noise represents – pollution, greenhouse gases, the unfettered exploitation of anything we can grab, the imperatives of economic growth – is entirely at odds, profoundly and irredeemably irreconcilable, with everything that Edward Thomas marvelled at on his hillside in Hampshire.

Edward Thomas didn't have to confront the agonies and dilemmas of climate change, obviously, and although he lamented the passing of a rural golden age, he also mocked the idea as absurd even as he wrote about it. He knew what his public wanted, though, and in his early writings every cottage is thatched and the cheery blacksmith's hammer startles the geese on a hundred village greens. But he wrote with unrivalled poetic detail in his notebooks about the natural world, of the 'ducks gathering insects from the surface of a pond at dusk – skimming them literally, with sharp snaps of their beaks, which they hardly dip'. Of the 'catkins swelling on the birches; they are silent now, but in May at their birth make a pattering against the stiff leaves'. And he wrote of the wildlife of London, of the 'hawthorn blossoming, in scattered sprays, on Wandsworth Common; while the leaves are gone, or going, with purple about them'. Thomas wrote about nature wherever he found it, because nature *is* everywhere. And the traffic, baying from the valley, is a reminder that there is no real point in only trying to save or preserve our woods and fields out here in the countryside (although we must), because we're going to have to do much more than that. Nothing that happens here can any longer be isolated from what goes on in Southampton, or London and Beijing, or the burning forests of Indonesia.

I wish I hadn't driven here.

The very top of the Ashford Hanger, known as Shoulder of Mutton Hill, is a wooded strip of yew and oak and beech, tumbling down both sides of the hill. Edward Thomas loved this place and wrote about it often, making it the setting for his poem 'The Path':

> Running along a bank, a parapet
> That saves from the precipitous wood below
> The level road, there is a path. It serves
>
> Children for looking down the long smooth steep,
> Between the legs of beech and yew, to where
> A fallen tree checks the sight: while men and women
> Content themselves with the road and what they see
> Over the bank, and what the children tell.

The path, winding like silver, trickles on,
Bordered and even invaded by thinnest moss
That tries to cover roots and crumbling chalk
With gold, olive, and emerald, but in vain.

FROM 'THE PATH'

Does nature writing always work better as poetry? Robert Frost obviously thought so – and he persuaded his friend Edward Thomas to make the attempt. For many of his poems, Thomas simply took his earlier prose pieces and notes and redrafted them into poetry. He had been 'a poet' all along, but it was only in the last couple of years of his life that he felt able – brave enough? – to embrace the form. He even published his first poems under a pseudonym, Edward Eastaway, as though he were trying to distance himself from what he considered years of hack journalism. Overall he published dozens of poems between 1914 and 1917, and the intensity of the poetry, and Thomas's growing confidence in his art, loops back into the prose. Now more than ever, rather than just noting the natural world, however exquisitely, Thomas makes his character and views an integral part of the writing. You could say he has something to answer for.

The children still play in the woods above Steep, that much is clear. There's a lean-to den here, made from coppiced ash poles and shaped like Eeyore's house in the Hundred Acre Wood. There's a frayed blue rope hanging high from a branch in a beech tree. Further on, there's a fallen yew, looking like it has hauled itself from the waterlogged ground and then sprawled down the hill, chalk crumbling from its exposed root base. I stare into the hole and think of Edward Thomas, lighting his pipe for the last time. The yew tree is mature, green, still growing, only now it is horizontal, its branches adjusting to the light and swerving skywards. I'm also thinking, as I head along the ridge, how very rarely I meet other people in the woods I walk through. The paths are well-trodden, but the woods are empty.

Just at that moment I see a middle-aged woman coming fast up the hill, strong and stooped under a heavy bag. She looks, like so many people you meet unexpectedly in woods, out of place and time, a wise woman gathering faggots in the fairy-tale forest, absurdly dressed in Gore-tex. I want to ask her the way to Cobbett's View, a place that the great polemicist William Cobbett raves about in one of his *Rural Rides*, but she charges past me at a pace I would struggle to match. It doesn't matter, as around the next corner is a Hampshire County Council sign, with this text written next to a picture of Cobbett, winking at us like a ruined old roué:

> Out we came, all in a moment, at the very edge of the hanger! And never in all my life was I so surprised and so delighted! I pulled up my horse; and sat and looked; and it was like looking down from the top of a castle down into the sea; except that the valley was land, not water.

And this view, too, can't have changed much in the past two hundred years, since Cobbett gazed out approvingly from the back of his horse. There are small fields and hedgerows, coppices and woods, the odd farm, but nothing built-up, the old field-patterns kept, nothing really to mark the passage of time other than the dull chunter of the traffic, muted now by a grey fog that is seeping over the land. We'll come back to Cobbett, but he liked a land with woods:

> Invariably have I observed, that the richer the soil... the more destitute of woods; that is to say, the more purely a corn country, the more miserable the labourers. The cause

is this, the great, the big bull frog grasps all. In this beautiful
island every inch of land is appropriated by the rich.

It's a simple formula. The woods around here have been spared
because the soil is too thin and the slopes are too steep for crops.
The great, big bull frog cannot be bothered to reach out for such
a meagre morsel: he's already gorging on the fat of the land.
Away from carefully conserved beauty spots such as the Ashford
Hanger, I don't think Cobbett or Edward Thomas would recognize
one hundredth of what they once knew and loved. The pace of
change has been unrelenting; and there is now more fecund and
varied growth in the back gardens of our cities than there is in the
chemical-drenched fields of 'the countryside'. The honey tastes
sweeter in Balham than it does in Steep. And what of the woods?
Cobbett is right: a land without woods is a destitute land, its people
impoverished, its soil depleted. Why, then, have we allowed our
country to lose so much? Who was meant to be looking after the
woods? Where did they all go?

3

'I speak for the trees, for the trees have no tongues.'

DR SEUSS,
FROM THE LORAX

The Word from the Woods

Is it possible to tell a history of the woods that puts them at the centre of their own story? History belongs to the victors and that's us (for now), but what do we think matters to the woods themselves? Specifically, what about the trees? How do they see things? It's a uniquely human question – we assume – and not one that is going to get an answer outside the pages of Tolkien. Even so, there are yew trees in Britain that are over three thousand years old, possibly much older. Imagine what they could tell us. So to hell with anthropomorphism: here's my list of the really Big Issues that must be preoccupying the trees:

- Their own survival.
- The survival of their species.
- Sex.
- Offspring.
- The avoidance of disease.
- The postponement of death.
- Sheep and deer (or where to find a wolf).
- Clean air.
- The right kind of soil.
- A benign climate.
- A world free from the wrong sort of caterpillar.

It's not very different, I would guess, from a list that a human might draw up, if you asked them to try and think beyond what they are having for dinner tonight.

There have been trees in Britain for millions of years. Of course, for the vast majority of that time there was no such thing as 'Britain' – it was either connected to the European mainland or buried under a mountain of ice. Sometimes both. There have been times when the land has been positively equatorial, with hippos munching on tropical palms. Around 700,000 years ago you would have seen early humans hiding from lions and cave bears in what is now East Anglia. And yet from about 180,000 to 70,000 years ago there were no humans at all (and, in all probability, no trees). What survived or died has been wholly dependent on the fluctuating climate, the glaciers advancing and retreating as the earth cooled and warmed.

The past 11,500 years have been unusually stable (up to now), and the descendants of the most recent band of human migrants to follow the retreating ice across the land bridge from northern Europe to Britain are still here today. There are no genetic remnants from the previous six attempts, but around eighty per cent of the people who live in modern Britain share DNA with this post-glacial wave of intrepid hunter-gatherers.

Over the next two thousand years the glaciers receded and the animals, humans and plants kept arriving from Europe. By 9,500 years ago most of the ice had disappeared from Britain. Released of its weight, the north tipped and Scotland rose. There are skeletons of whales buried in the Scottish hills. Trees were thriving in the warmer climate – juniper, birch and Scots pine, alder and oak – and steppe animals such as reindeer and wild horses left for Scandinavia or died. Britain was becoming a forest and other animals were flourishing under the tree cover: the wolf, wild boar, roe deer, elk and brown bear. As the glaciers melted, the seas kept rising and whole forests died standing upright in the rising tides. The land passage across the Straits of Dover disappeared and then, about seven thousand years ago (*about*; academics are still arguing and who was counting?), the last land route between the British Isles and Europe, somewhere near the Dogger Bank, became an impassable swamp and then sank under the waves. Britain had become an island. It seems, under the circumstances, very *recent*.

In this chapter, a brief history of the woods of the British Isles, I have tried to make the woods and trees the stars of their own story. I wanted to do this without imposing my own human perspective, and so, in the section that follows, I have made every word correspond to the passing of two years, regardless of what we might think of as important. I emphasize events that affected the woods, not us. I have assumed that Britain became an island in 5,000 BC, and because at the time of writing the year is 2018, this history is exactly 3,509 words long. It's not an infallible system, but the first word of this history starts in the year 5,000 BC and you will find, if you check, that the 2,521st word takes you to the arrival of the Romans, carrying their rabbits and sweet chestnut trees; and the 3,033rd word ushers in William the Conqueror. Events escalate after that – and although I've tried to place the woods centre stage, in the end this story became all about *us*.

A SHORT HISTORY OF BRITAIN'S WOODS IN 3,509 WORDS

There's a final rush to beat the rising seas. The last tree to reach dry land in the newly minted British Isles is quite possibly the box. There are, perhaps, twenty-six species of tree now marooned on these islands. Juniper was here early, and has already spread to the far north of Scotland, followed by the silver birch and the Scots pine. There are alder trees along the rivers, lakes and streams. In the south of the country the most common tree has become the small-leaved lime, basking in the warmer climate, but there are also great stretches of oak, ash, hazel and elm. There are beech and hornbeam settled in the south-west. Further north, the oak is dominant, but the hazel is also here in numbers, its nuts spread by red squirrels and birds. The pine and the birch are retreating ever northwards, only surviving further south in the scraps and edges of the broadleaved forest as the shade deepens over the young land. The first oak reaches the Firth of Forth. Rowan and willow, aspen, poplar and the late-arriving hawthorn make their way north and west. Human hunter-gatherers follow through the glades and make temporary clearings in the woods; perhaps they favour and spread the hazel, wild pear, the crab apple and the cherry; there are brown bear, lynx, wolf, deer, bison, elk and boar in the forest; beavers dam the rivers and fell the trees. This is the wildwood, a place of constant change, a kingdom of trees. A yew berry passes through the gut of a jay and is excreted by the side of a stream in Fortingall, Perthshire, the start of its millennia-long life journey.

Humans have discovered arable farming, animal husbandry and pottery, or it has arrived from Europe. There are pigs in the woods and cattle on the fringes. There's a 'Neolithic Revolution' under way and land is being cleared, very slowly, but the tree cover is expanding. There might be ten thousand humans in Britain, there might be a hundred thousand, no one knows. They are almost certainly outnumbered by wolves. The maple and ash are thriving on the calcareous hills; black poplar and holly have followed the birch to almost every part of the land. Wheat is being planted in the south of England. Trees are spreading up the mountains, mostly juniper, rowan, holly and birch, although there are tracts of bare ground at the higher levels. Wild aurochs browse in the forest, tearing at the saplings. The aspen trembles in the cold north-western isles. An elder tree flowers and fruits in the far south-west and the first elm tree has reached Scotland.

Some of the humans have stopped roaming and are perfecting the art of farming. They still hunt, but they come home. More wheat is planted. The juniper has colonized the last mountain, growing at six hundred metres on the slopes of Beinn Eighe. Blackthorn and bird cherry blossom drift through the woodlands in springtime. The forests are dense with fallen trees and the only paths are made by elks, aurochs and bison crashing through the undergrowth and browsing on the leaves, or by the predators that follow them: wolf, lynx and human. It is a shifting patchwork of deep wood and some more open areas, kept clear by deer and, at the fringes, by humans. More of the trees are being felled and they are now stopped from creeping back into the fields by grazing domesticated animals and diligent farmers. The elm tree starts its long retreat, weakened by a changing climate, possibly by disease, its leaves stripped by humans for their animals' winter fodder. The small-leaved lime reaches its furthest point north, somewhere in the Lake District.

Humans have laid a track of oak planks, supported by elm, ash and lime posts, through the marshes of Somerset. Some woods are being coppiced and managed for the first time. There are villages with wooden buildings, temples and tombs. In the north, where the climate is still notably colder, the Scots pine is dominant, with rowan, birch and heather growing between the trees. Wild boar, rooting and truffling in the forests, create glades for wildflowers and the insects that feed on them. The trees rot where they fall, providing crumbly nourishment for millions of beetles and the predator woodland birds. Nightingales sing in the clearings. The human population is growing, agriculture is becoming more sophisticated and the grassland is spreading, and while most people still choose to live along the coast and the more open habitats by the sides of the rivers, they are also forging deeper into the woods. They chip away at the edges of the wildwood, make clearings where they can, but they avoid its depths.

There are colonies of walrus on the seashore and vast schools of grey whale in the teeming seas. Pelicans squabble with puffins. The great auk builds its nests among the distant rocks, although humans have found that its flesh makes a tasty meal and its down the softest of beds. Oak, birch, juniper and pine fill every last crevice of the coastline, while the glades and the fields widen gradually behind them. Perhaps we can now say that humans have replaced climate change as the main factor affecting the woodlands of Britain; in other parts of the world you could even say that humans are starting to affect the world's climate for the first time, as they torch their forests. The woods of Britain are too wet to burn. Sometime around now we arrive at 'Peak Forest': the trees are everywhere, but the tide is turning. People must also be just about everywhere – or is it wolves? – because the European bison becomes extinct in Britain. In Scotland, the trees are being cleared, but are converting to moorland. And in the far damp north-west the peat bogs are slowly spreading.

The use of coppicing in woodland is intensifying. People are managing the trees, making sure they have enough of the right species and plenty of timber of the right size for their needs. Long, regular-shaped poles are being used for homes and fences, for boats, buildings and fires; and huge timber piles are used to transport bluestones from Wales to Stonehenge in Wiltshire – you couldn't do this through dense forest. The leaves and bark of the lime and the elm are being stripped for food. In fact, the elm is in full retreat, some of the gaps in the forest being filled by hazel and ash, but much more of it filled by people and their animals and farms. It is the beginning of the end of the wildwood. There are more and wider clearings for agriculture, and the moors are spreading that once were forest; there are significant breaks in the tree cover – deer and sheep keep it that way. A red squirrel can no longer bound through the treetops from coast to coast, if it ever could. But the woods are still vast and abundant in most parts of the country and they are dense with wildlife. Most of the indigenous large mammals – aurochs, wolves, bears and elks – still thrive. Freed from the summer shade, there is a blooming of flowers in the new glades, meadows and fields. It is good news for the honeybees. The beech tree and the hornbeam have spread from their south-west homeland and are creeping slowly east and north. The hazel is faltering, unable to pollinate in the shade of the larger trees.

We have reached the Bronze Age. New technologies, such as the metal plough and axe, are speeding up the transformation of the landscape. Trees can be chopped down more quickly, their roots grubbed out, and the return of saplings prevented. The human population is growing – as is the number of sheep. Wolves are retreating deeper into the woods and mountains. Wood is now being used as fuel for industry, for the smelting of metals and the firing of kilns. In any case, the woods are (as ever) in their own state of flux; oak, lime and ash trees are abundant in the south; pine and birch still dominate the north. Nets with poles are cast across the rivers and the number of wooden boats multiplies.

Trees are useful to humans, but they are also worshipped. There are holy trees and priests and sacred groves in the forests, where animals and humans are sacrificed. Perhaps trees are seen as links with other worlds, their roots reaching down to the underworld, their branches stretching high to commune with the sun and moon deities. Nature is bountiful and the supply of trees must seem eternal and without limit; at least, we do not think they are being coppiced for reasons of conservation, but convenience. A huge oak tree is buried in its entirety, upside down, near what is now Holme-next-the-Sea in Norfolk. And somewhere in the wildwood the last auroch, Europe's massive, indigenous wild ox, comes to a lonely end.

People use dogs to hunt in the woods, but the best hunting is in the glades and clearings. This is where the roe deer gather, as well as the dwindling populations of elk. Beavers are still creating dams for pools and disrupting rivers by felling trees, but they are also being hunted for their fur. The wild is in retreat. There is an accelerated contraction of the wildwood and the open grasslands are spreading deeper inland. Wood pasture and pollarded trees are further changing the landscape, but the greatest change is the intensification of agriculture. The new technologies have disrupted the human relationship with nature: there is no part of these islands that is not now, or could not be, exploited by humankind. The human population is expanding, but it is well fed by land and sea. To keep growing, though, it needs more land. There is trade in wooden boats with mainland Europe. Around about now, as a rough estimate, about fifty per cent of Britain is still covered in woodland, much of it thick with tangled undergrowth and inaccessible to people. Tin mines are being opened in Cornwall and there is a pressing local need for timber, to make the pit props and feed the forges.

The main hills in England and Wales are being cleared of tree cover, ditches are being dug and wooden ramparts erected. The demand for timber is growing: it is plentiful, although the Scots pine, such a useful straight tree for constructing buildings and boats, has almost entirely disappeared from England. It is also the only British tree to burn easily in the damp climate, and it cannot sucker or coppice, which makes it easier to eradicate. The beech tree is starting to dominate some southern woodlands, creeping into small clearings before overshadowing its competitors. This is good news for pigs, which feast noisily on the delicious nuts. Humans and their animals are roaming deeper into the woods, the paths and clearings are being widened, the undergrowth is thinning. It is a land of grasses and moors, separating a patchwork of small and great woods. There are now substantial population centres across the British Isles, especially in the south and west, and there are hundreds of hill forts. Grassland flowers are flourishing – speedwell, buttercup, orchid and gentian – now that there is so much less competition from the trees. The elm is no longer one of the dominant British trees, but it will return. The elk, on the other hand, will not: it has been hunted to extinction, as has the last British walrus. The woods are quieter, and so, too, is the shoreline.

The climate is changing and becoming wetter, forcing many people from the hills. Ever bigger farms are developed for live-stock in the lowlands, boosting demand for timber and land. The best land – where the lime trees once sunk their roots and grew in such abundance – is now almost entirely cleared and taken up with agriculture. Humans are fighting each other with increasingly sophisticated bronze weaponry, and it's possible there was a major invasion (or migration) into southern Britain at this time. The country is certainly more crowded. Places where the soil is too poor for a primitive plough to have much effect (the clay of Kent and Sussex) or where the land is hard to reach (the mountains of Wales and the north) remain heavily forested. Dartmoor, which was once a forest, and then farmland, is becoming a moor. The human population continues to grow; it is the other large, undomesticated land mammals that should have cause for concern.

There is another dramatic technological leap forward and humanity enters the Iron Age. As well as advanced weapons and sharper axe blades, this brings the iron plough into Britain's woods and fields. The heaviest clay is no longer off limits. Roots are grubbed up more rapidly and without the need to keep changing the blades. Vast new areas are now under cultivation and the small-leaved lime almost entirely disappears from England, ravaged by humans and the changing climate. Industry is spreading (tin and copper mines, the production of iron) and the first coins are being minted. The demand for trees and timber – for fuel, buildings and fodder – grows greater; and the management of woods (with coppicing and even fencing to repel deer or livestock) is now a widespread necessity. The human population of Britain stands at about one million.

The first towns are being built, encircled with wooden walls and using large quantities of timber for their construction. There are wooden buildings, wagons, fences, pathways, boats, tools, beds and barrels; while the new wooden lathes are being used to turn out the spokes and hubs for wheels, buckets, handles and bowls. Above all, wood is a fuel, keeping people warm in the winter and filling the blazing furnaces of their industries. It is the Iron Age, but it is a world of wood. All around Britain, the forests are falling. Trade between the peoples of Britain and Europe is complex and vigorous; there is a need for strong timber for boats, which are also used for fishing. At about this time, the last British grey whale is hunted to extinction. Back in the woods, the Druids tend their sacred groves and pick mistletoe in the moonlight. The yew and the oak are worshipped for their power and longevity. As the woods fall, a number of other plant species benefit. Primroses, bluebells and foxgloves bloom in the clearings as the woodcutters follow their cycle of coppicing. The sheep browse happily in the pasture underneath the pollarded trees. The wolves are far away.

The peoples of the Mediterranean, who have already deforested their own hillsides and dug out their ore, have heard of Britain, with its wild blue tribes, tin mines, slaves and plentiful forests. They send an expedition and they land on the south coast. Briefly. They see a thickly wooded land and unfriendly natives. On the eve of the Roman invasion, Britain has perhaps thirty to forty per cent tree cover, but in parts of the country it is much more. Ninety years after their first visit, the Romans are back to stay. They do not transform the landscape, but they intensify its use. The sacred groves of Anglesey are destroyed, the Druids killed. London is founded. Many forests are cut down, making sure the natives have nowhere to hide. Hadrian's Wall is built across the north; the Scottish woods and people are left to their own devices. New towns are built, needing unprecedented quantities of timber. The sweet chestnut arrives. So does the small Mediterranean rabbit, a keen consumer of saplings. There is a boom in iron smelting in the south and east. Charcoal is burnt and more woods are tightly managed. New farms spread across the land and the plough does overtime. The sweet chestnut is planted in Kent and Sussex. It is no good in this climate as a nut, but it is exceptionally useful for coppicing. There are towns; and there is countryside. And then there are the woods and wilds, much of it now confined to the far north and Wales. The Romans start to leave Britain. The ash and the beech are thriving; the hornbeam is settling into its hinterland around London.

The Romans have gone, although people are still working in the fields and woods and many villas and farms remain occupied; there is just a general drift of decay, into which the Angles, Saxons and other Germanic invaders arrive. Darkness falls on the land and Arthur rides through the wildwood. The last lynx is killed. The towns empty of people and the trees return. The woods are expanding, although this doesn't last long; the Saxons are keen and proficient farmers and have many uses for wood. What remains of the wildwood is left largely intact and used for hunting. About twenty-five

per cent of the land is under the cover of trees, much of it as thick
as it ever was. The Saxons tighten their grip across England. Every
inch of this new land now has an 'owner', even its woods. As the
fighting dies down, the land is worked more intensively. Only the
woods of Wales and Scotland seem to fare better, or at least they
are left relatively undisturbed. The iron-smelting industry eradicates
the Forest of Arden. The towns expand, but this is a rural society
and the open fields are kept free from invasive trees. Vikings from
Scandinavia arrive looking for land and timber for their ships. It is
a time of fighting and the woods flourish. The wolves, though, are
still hunted intensively.

King Alfred builds England's first navy, using the native oak
trees. An idea is born that English oak is better than any other and
the species begins to be favoured by foresters. Some time around
now, Britain's last brown bear dies. The population of humans is
growing again, boosted by another wave of land being brought
into agricultural production. There are perhaps 1.5 million people
living in Britain, most of them working the land. They are healthy
and prosperous, living in a wooded agricultural land with a few
scattered towns and many sheep. The Normans arrive. Rufus dies
in the New Forest. The kings assert their ownership of the forests
of Britain and Ireland. Big European rabbits have arrived. There
is frantic building activity: towns, roads and castles. Coppicing
spreads and wood banks keep the forests free from livestock and
deer. The last wild boar is hunted and killed. There is fifteen per
cent woodland in Britain – just above modern times. The Weald
and Chilterns have seventy per cent cover. The woods are full of
people and their pigs, claiming pannage.

Wood pasture increases. The sycamore arrives in Britain and
spreads with vigorous ease. The woods are being felled. More
heaths and moors are created on land with poor soil, the hungry

livestock preventing the return of trees. Sheep numbers are soaring. There are now nearly five million people living in Britain. Then Black Death strikes. The population falls to 2.6 million. The woods return, briefly, but the amount of wooded land in Britain is just seven per cent. The walnut tree arrives. The human population continues to fall. There is land to spare, although many forests are the preserve of the king and his hunting parties. People are leaving the lord to his manor and setting up on their own. The spruce is brought to Britain.

There is a huge increase in iron smelting, leading to the deaths of tens of thousands of oak trees. Plane tree arrives. Last beaver killed. Market economy. Industry thriving in the woods. Landed gentry making fortunes selling timber. A bigger navy needed. Arrival of fir trees, horse chestnut, false acacia and the first tree plantations. First Acts of Enclosure, keeping people out of the forests. The larch arrives and is deemed to be the perfect plantation tree. Boom time in the iron-smelting industry; more coppicing; more forests felled. Last wolf killed.

The navy expands rapidly, needing timber. English oak is favoured. More timber imported. Great expansion in farming – woods grubbed up and corralled into smaller areas. Landscape gardeners move trees and uproot woods. Larch and sheep prolif- erate. More farms. Human population now six million. James Watt patents a steam engine with a continuous rotary motion ('nature can be conquered, if we can but find her weak side'). Coalfields. Wood no longer primary industrial fuel, but consumed for railways, mines, building. Drift and then flight to cities.

It's boom time for agriculture. Over one-quarter of Britain's remaining ancient woodlands are destroyed. Sheep, wheat, orchards. Slump in agriculture. Woods return, but not the same. South Wales woodlands destroyed. Cities. More plantations than ancient woods.

Forestry Commission. The Oak Change. Woodland cover under six per cent. The Locust Years and destruction of over half of last ancient woods. Conifers. Motorways. Woodland Trust. Dutch elm disease. Deer. Ecology movements. Storm. World warming. Acid rain. Ash dieback. Conifers felled. Human population sixty million. Beavers return. Floods. Rewilding.

4

Give me a land of boughs in leaf,
A land of trees that stand;
Where trees are fallen, there is grief;
I love no leafless land.

A.E. HOUSMAN,
FROM 'GIVE ME A LAND OF BOUGHS IN LEAF'

Until You Were Gone

The world spins faster and perhaps it's better not to get too attached to those returning beavers. A report on the BBC website on 31 January 2016 had this mournful news: 'Beavers that were heavily pregnant or had recently given birth are among those shot by landowners in Tayside. The news has led to demands for restrictions on shooting during the breeding season and renewed calls for Scotland's beavers to receive legal protection. The country has two beaver populations, despite the species being hunted to extinction in the 16th century. The Scottish government said it was taking time to consider the issue.' Apparently the 'landowners' were upset by the beavers' tendency to gnaw down trees, build dams and create small ponds. No one stopped to ask the beavers what they might have thought about this high-handed and terminal intervention. If they had, the beavers might quite reasonably have replied that they were working busily in the woods upstream to slow and spread some otherwise dangerously trammelled rivers; that they do in fact run a complex and highly effective flood control system; and in any case they were only minding their own business while trying to raise their families, preferably as far away as possible from a seething horde of murderous primates with an overdeveloped God complex. Isn't it rather a strange idea, they might have added (as the lead began to fly), that one dominant individual from a single species can 'own' the land? And even if that were possible… *we were here first!*

Miraculously, at the end of 2016 the Scottish government came to the end of its deliberations and announced that the beaver was indeed a native British species and that it should be treated – and protected – accordingly. This was joyful news… and surprising to many. But beavers lived in Britain for millennia and it is us that had forgotten: so thoroughly have we excised them from the landscape that most of us think of them as alien, exotic and exclusively *Canadian* creatures. If we imagine them at all, they are toppling pine trees in the vast, snow-gripped woods of the north. Once upon a time they also lolloped through the forests of our childhood dreams. I learned everything I needed to know about beavers from *The Lion, The Witch and The Wardrobe*, the first book that C.S. Lewis wrote in his Narnia series. They are on the side of Right. Their whiskers tickle when you lean in close to hear a whispered plan. And if you have been struggling through the frozen woods, lost and far from home, then you could not ask for a better, more exuberant welcome than the one given by Mrs Beaver:

> 'So you've come at last!' she said, holding out both her wrinkled old paws. 'At last! To think that ever I should live to see this day! The potatoes are boiling and the kettle's singing and I daresay, Mr Beaver, you'll get us some fish.'

Like Tolkien, his fellow children's book author, elf-obsessive and drinking partner, C.S. Lewis despaired of what the twentieth century was doing to the land and the woods he loved. In *The Lion, The Witch and The Wardrobe* one of the four children, Edmund, escapes from the others in order to betray them to the evil White Witch. As he stumbles through the forest, he broods on what he'll do when he's king, 'what sort of palace he would have and how many cars and all about his private cinema and where the principal railways would run and what laws he would make against beavers and dams'. All the fanciful embellishments of the twentieth century will be visited upon the land. But, by the end of the book, Edmund has grown and been transformed into the very best kind of king. With his brother and sisters, he 'made good laws and kept the peace and saved good trees from being unnecessarily cut down, and liberated young dwarfs and satyrs from being sent to school, and generally stopped busybodies and interferers and encouraged ordinary people who wanted to live and let live'. The woods are saved, the rivers run clean, and eternal winter (with its threat of cars, railways and private cinemas) is banished from the purified land. The beavers and the rest of the animals live happily ever after, fawningly grateful to the four humans (without whom, we are left in no doubt, the world would have gone to hell).

Lewis could get away with this childishly simple message because he had written a children's story (he'd be accused of peddling arrant nonsense, but it's surely slightly undignified to get too agitated about a book that features a talking faun). He has of course crowbarred a Christian, feudal, paternalist agenda into his simple tale, but it's still a beautiful, evocative story, and its real power lies in the descriptions of a devastated land being healed by the power of love. The children may be in charge, but in reality they are only first among equals in a world of talking animals and trees, and they walk at their ease through an earthly Eden. It's a wonderful place, Narnia; and its woods are places of refuge and harmony.

As the pace of industrialization (and the horrors of industrialized warfare) escalated, Lewis wasn't the only writer to seek sanctuary in the world of children's writing. Tolkien, of course, but also A.A. Milne, Kenneth Grahame, Rudyard Kipling, Arthur Ransome, T.H. White: it was no doubt easier to slip away into an unsullied world of sun-dappled meadows and to pretend that if something diabolical was happening, it was all happening somewhere else.

The Edwardian men seem to have been especially upset. And why not? The changes they were witnessing, the uprooting of rural life, the felling of the forests, the catastrophic belief that we could uncouple ourselves from nature, were all real enough. E.M. Forster was traumatized by what was happening:

> The growth of the population and the applications of science have destroyed [England] between them. There was a freshness and an out-of-door wildness in those days which the present generation cannot imagine. I am glad to have known our countryside before its roads were too dangerous to walk on and its rivers too dirty to bathe in, before its butterflies and wild flowers were decimated by arsenical spray, before Shakespeare's Avon frothed with detergents and the fish floated belly-up in the Cam.

Forster's lament to the death of Old England was written in 1960 in an updated Preface to his novel *The Longest Journey*, but he is talking about the year 1907, the time when the book was first published. Long before that second Preface, though, he'd given up writing novels. Under the circumstances, and convinced that his land was dying, he didn't feel he had anything worth saying: 'Peace has been lost on the earth and only lives outside it, in places where my imagination has been trained not to follow.'

There are problems for anyone who mourns the passing of an emptier, wilder time, the main one being that we find it very hard to trust our own reactions to what we can see happening around us. If a patch of wilderness disappears, or a lake is drained, to make way for new roads, crops, or conifer plantations, or if fly-tipped fridges are sprawling down the banks of a much-loved wood, most of us find it possible, after a period of grief, to accept the changed

landscape. The changes are incremental – even if they are relentless. And maybe it's possible that they're not even so bad. After all,
perhaps we could say that most things are now just that little bit
tidier. And if there are changes, they are merely signs of the progress
and growth we have been taught to expect and welcome. We have
become muddled about what we truly value. For centuries, almost
no one would have dreamed to say that a new road was ugly, noisy
and toxic, and certainly not unnecessary – it was an essential step
forward into a better, speedier future. What kind of retrograde nostalgic would want to stand in its way? Poets and losers, for the most
part. All the protesting Edwardian men had to offer in return was
a wistful whiffling about moles, riverbanks and heffalumps, while
the future was going to bring us cars (for everyone!), silver rockets
and a nuclear dawn. Come to think of it, wasn't there something
suspicious about those early, anti-progress protesters? Isn't it a fact
that they just wanted to keep everything for themselves and their
tweed-clad chums?

Edward Thomas seems to have been just as confused as the
rest of us. He wrote with heartfelt urgency about the disappearing
meadows and dwindling beechwoods, but could announce in the
next sentence that his Edwardian contemporaries were fighting
to preserve something that was never there in the first place. He
treasured the woods and fields, and thought that the towns had
'perhaps' grown too big, but he also suspected that people's selective
memories of their golden childhoods had somehow become entangled with a glorious but illusory golden age – and that neither thing
had ever really existed: 'We blink, deliberately or not, unpleasant
facts in our own lives, as in the social life of [Ancient…] Greece or
the Middle Ages.' He thought 'the happiest childhoods are those
which pass completely away and leave whole tracts of years without
a memory'. He basked in the fading golden glow of an Edwardian

rural idyll – you could say that he helped to create it – and then he dismissed the whole.thing as an illusory dream.

I'm in a fretful mood as I head back to the scene of my own childhood. I spent the first fourteen years of my life in a village called Wadhurst in East Sussex. 'Wada's Wooded Height' I was taught it meant – and that's what it was, even then: high and wooded, despite being on the very outer reaches of commuterland. The nearest town is Tunbridge Wells and it strikes me, as my car crawls down the busy streets, that even now this is a place (as was Wadhurst) that has been infused and shaped by its surrounding woods. There are trees everywhere: on the pavements and looming hugely in the many parks; there are even ancient coppiced woods pressed up against the outskirts of the town. Housing developments, retail and business parks have taken great chunks out of the woods, road-widening schemes have stripped out more, but this is still a town where you can feel the distant breath of the forest, where it's possible to imagine that the town is a clearing in the woods, and not that the woods are beleaguered scraps in a murky sea of human development. Both Tunbridge Wells and Wadhurst were carved out of The Weald, and even now this is one of the most heavily wooded parts of Britain. The name, 'Weald', means forest in Old German and I am driving through what was once the old Saxon Forest of Andred.

It is a long time since I was last here and I find I am peering anxiously through the car windscreen looking for signs of change. No doubt I am also trying to swipe away a golden fog of memories. I do know that I won't be able to find the house where I grew up: it was knocked down in the 1980s, a few years after we'd left, to make way for a new housing estate – which is the main reason I haven't felt able to stop here since. The house was built in the 1930s and was no great loss to anyone (other than us), certainly not for architectural reasons, even if it did have its very own air-raid shelter. At the time it stood on the corner of a crossroads on the main route into Wadhurst, just across the road from the sweet shop, its entrance next to a Shell garage. (The man who worked in the kiosk there gave us free football cards and I'd filled my 1970 World Cup album long before anyone else, apart from my more obsessively completist brother. I still have two books filled with 3D stickers of endangered animals, including a polar bear and her cute cub. Thank you, *Shell*, and sing ho! for the icy north.)

Our home was set far back from the road at the end of a circular driveway. There were once oak trees ringing the perimeter and I'm relieved to see they are still there. The name of the house was 'Little Park' and it stood in just over three acres of land, which my parents had spent twenty years turning into their very own garden paradise. They dug, built and tended everything themselves, helped by Mr Longley, a short, military-straight, brilliantine-haired veteran of the First World War, who spoke with a lulling Sussex burr and wore a heavy three-piece suit while gardening, in all weathers. He'd discuss the war and the trees with my father and the disastrous state of 'them beans' with my mother; and every year he would lament something he called the 'blossom wind' that carried away the orchard flowers before they'd had a chance to shine. He spun elaborate nets of thread on the cherry trees in a doomed attempt to deter the ravenous bullfinches, which he regarded with loathing, and I sat in his wheelbarrow and unwrapped sticky rectangles of Callard & Bowser toffees that he fished from his waistcoat pocket. Mr Longley and my parents created dry stone walls, stepped lawns with great drifts of daffodils, herbaceous borders brimming with phlox, snap dragons, poppies and pansies, greenhouses heady with cucumbers and tomatoes, vegetable beds for asparagus, strawberries, potatoes and beans; a tumbling orchard of cherries, apples, plums and pears; a pond stocked with goldfish (although it was soon only good for newts, dragonflies and water beetles, whose larvae ate the fish); a copse of maple and acer, banks of rhododendrons, a towering cedar for climbing; and, at the centre of it all, a fenced rose garden that was my father's pride and joy and in which he would wander in the early mornings, searching for the perfect fresh rosebud to wear in his buttonhole to work, or linger late on summer evenings, breathing the roses and supping on a tumbler of Scotch. We played cricket on the lawns and raced our bicycles down the paths. There was even a small wood, which for some reason it must have amused my parents to call 'The Spinney', and we loitered here for hours under the shadow of the steepling trees, scavenging for chestnuts, tending the bonfire and building dens from the fallen branches. I learned to love nature here, without even thinking about it, especially the trees. I had no idea how lucky I was.

All of this has now gone. There is a red-brick car park where our house once stood, with fifty-five numbered spaces for the rows of cars. There's a rank of scarlet wheelie bins, and pavements and streets where the lawns once grew. And there are houses, of course, in a modern Sussex style, nodding back to the old cottage vernacular (an improvement, you might say, on my old home). The whole development looks solid, and neat, even on this ragged November day, with the leaves drifting into the gutters and hedges. I am searching for something I might remember. I knew this place, these slopes and this sky, more intensely than anything I've known since, but it's as though everything has been hosed away and buried under layers of brick. I look for the trees, but the cedar has gone. Perhaps the sky seems familiar, but that is surely not possible. I do, maybe, recognize the treetops across the road – or at least I think I do – and is that a remnant of The Spinney behind one of the houses? But it has been almost forty years, and I can't find what

I'm looking for. Even the horizon has changed: the horses have gone and there are more new houses in the field that once backed onto our garden.

It is midday and nothing moves. I don't feel very welcome. I probably shouldn't be here at all, brooding about change and peering into other people's houses. But 'they've paved paradise and put up a parking lot'. *My* paradise. Ghosts flicker and whisper just out of sight. And birds sing. In fact, once I notice them, I realize there are hundreds of birds here, sparrows, blackbirds, finches, blue tits and robins, and they are singing their hearts out. Are these the descendants of Mr Longley's bullfinches come to say hello? It's an absurd thought and, honestly, I need to forget it. It's gone. All of it. There are new houses and new people here, just as my own house was once built on fields and, much further back in time, the Saxon Wada cleared the virgin forest to make way for their homes and pigs. We march on, trailing ghosts.

In George Orwell's 1939 novel *Coming up for Air*, the hero, George Bowling, heads back to his idyllic childhood village in Oxfordshire, and finds that it has not been demolished, but 'swallowed': 'All I could see was an enormous river of brand-new houses which flowed along the valley in both directions... Queer! You can't imagine how queer! All the way down the hill I was seeing ghosts, chiefly the ghosts of hedges and trees and cows. It was as if I was looking at two worlds at once.' This is all we know. Everything changes; and we live with ghosts.

But enough of all that. I have strayed a long way from the woods.

When I was thirteen, my family moved seventeen miles to the east and into Kent, from Wadhurst to Tenterden. If we'd made the journey in Saxon times, we'd have walked the entire way through the Forest of Andred – 'Tenterden' means a clearing in the forest (a 'den') belonging to the people of Thanet. Now the woods are few and scattered, but they are still a presence. William Cobbett made his way to Tenterden in late August, and was much taken by the beauty of the local girls. He seemed especially thrilled with the young, rosy-cheeked lasses in the Methodist Meeting-house, and in particular their singing: 'We may talk of sparkling eyes and snowy bosoms as long as we please [hurrah!]; but what are these with a *croaking, masculine* voice? The parson seemed to be fully aware of the importance of this part of the "*service*". The subject of his hymn was something about *love*; and the parson read, or gave out, the verses, in a singularly *soft* and *sighing* voice, with his head on one side, and giving it rather a swing. I am satisfied, that the singing forms great part of the *attraction*. Young girls like to sing; and young men like to hear them. Nay, old ones too.' I mention this because we'd moved into an old Methodist Meeting-house in Tenterden. I have no idea if it was the same one – none of us had read Cobbett at the time – but I now like to think the place was haunted by the old bounder, ogling the girls and rubbing his thighs in leery appreciation.

One of Cobbett's preoccupations as he rode around the Home Counties was the government's claim that the population of Britain was rising. He refused to believe it, repeatedly returning to the fact that there were huge churches in villages across Kent, Sussex and Surrey that were only half-filled with people. He took this to mean that the population was falling; although the truth was that the rural poor were heading to the cities. Over 150 years later and London (Cobbett also calls it the 'great wen') has grown exponentially, while Tenterden heaves at its bounds. It's still recognizably the same place, though, even if the high street is clogged with four-wheel drives and the pavements are banked high with drifting clouds of pensioners. The small oak wood just off the high street, where my mother and her dog used to listen to the nightingales singing, has been converted into a supermarket car park, but otherwise things have changed less than I expected. At least our house is still there.

We used to walk in Knockwood, a mixed woodland – mainly oak and sweet chestnut – about a mile from our home. I'm back there now, in late summer, parked (appropriately) in Summer Close. The leaves are on the turn and there's a sound of chainsaws in the air, although not from the wood itself. I remember my father telling me, on this path that leads past the last house and into the murk of the woods, that at his age he had taken to 'spring hopping'; that he didn't know how many more springs he was going to live to see, so he very much wanted to make the most of this one. I was probably about sixteen when he shared this unwelcome insight, and teenage-surly, so I no doubt grunted him away. He was a gloomily romantic sort of man, and the family teased him that he liked to look for (and usually find) the worst in most situations. It seemed to give him a glum satisfaction, and he was easily moved to tears. He was already fifty years old when I was born – he was born in 1912 – and he was by nature, even when younger, old-fashioned in his ways. In fact, he was probably rather Edwardian, right down to his surprisingly natty dress sense. But I don't want to give the wrong impression. He had a deep twinkle in his eyes and he kept himself interested in the world and its people.

His favourite poets were Rudyard Kipling and A.E. Housman – although he preferred history books about Napoleon and the American Civil War – and this is probably where he got his notion of 'Spring Hopping':

> Loveliest of trees, the cherry now
> Is hung with bloom along the bough,
> And stands about the woodland ride

Wearing white for Eastertide.
Now of my threescore years and ten,
Twenty will not come again,
And take from seventy springs a score,
It only leaves me fifty more.
And since to look at things in bloom
Fifty springs are little room,
About the woodlands I will go
To see the cherry hung with snow.

HOUSMAN, *FROM* 'A SHROPSHIRE LAD'

My father must have been a few years short of his 'threescore years and ten' when he told me how he numbered and treasured each remaining year. It turned out that he had just about another ten springs ahead of him. His enthusiasm for life was not always apparent, but it was there, even if it was too often buried deep under layers of self-effacement. He'd had a successful (indeed at times thrilling) working life, punctuated by war – in the 1970s they'd have probably called him a 'Captain of Industry' – but that would have been startling news to his children, as well as to himself. He'd practically retired by the time we reached Tenterden. He'd disappear occasionally up to London with burnished shoes and a rose in his buttonhole (he'd built another rose garden by then), but at home he was quiet, amused, eager to play cards, drinking (although never noticeably drunk), softly rumbling about cricket, his own-cured hams and French wine. In his pomp he had stood six feet three inches tall, with the muscles to match, but he hated to stand out. I only once saw him lose his temper: my economist brother couldn't see why we needed to buy British cars, but after a stifled outburst ('I *love* my country, and that's it') he hurried to his rose garden. 'Setting all modesty aside' was how he would sometimes begin his sentences (and how we scoffed), but he never would. He was too shy, too much of an English father, to get to know really well; and that's painful to consider.

So I think it's because of him that my head is abuzz with Edwardians, as I make my journey through the English woods. My father felt a deep nostalgic connection with the English countryside (even though – or because – he'd spent half his childhood in Brittany and much of his adult life in the Far East). He would potter around old farms, poking at rusting machinery; and in Knockwood he would make a beeline for the place where he hoped to find the charcoal burners at work. Knockwood was energetically coppiced in those days and carefully chosen parcels of the woods were cropped on a fluid seven-year cycle. This was mainly sweet chestnut (for the charcoal), but there were also massive old stools of hazel and, I think, some alder. Certainly there was a sludgy stream oozing through the valley at the heart of the wood, and the alder likes its roots moist. By the mid-1970s there must have been almost no charcoal burners left in Britain, but weirdly there were a couple of old men here, in tiny Knockwood, working their trade – and following their own fathers' tradition in The Weald and beyond. I can remember standing on the edge of the clearing, more embarrassed than interested, while my father chatted with the charcoal burners, plying them with questions, stooping over the smouldering domes, snuffling his large nose at the smoke, nudging his size-thirteen shoe at any dead embers. There would be three or four heaped piles at any one time, coppiced chestnut that was layered with turf and then burned for a week or so while the wood turned ever so slowly into charcoal. The men built and lived in a hut in the woods while this was going on – they couldn't leave the charcoal unattended in case it burned too quickly – and then disappeared as mysteriously as they had arrived. My father was in heaven.

Knockwood doesn't seem to have changed, but there's no sign of any coppicing (and absolutely no charcoal burners). I may have caught it between cycles, but a wood that's not worked is, for much of its ecosystem, a wood that is dying. 'If you want wood, you have to cut wood' is the saying; and a tree that is regularly cut can live for millennia – far longer than a tree that is left alone. Countless species have evolved to take advantage of clearings in the woods – or perhaps it's better to say that the species that dislike dense shade have flourished in the years that humanity has been coppicing – and without the clearings we are losing many beloved plants and animals. Primroses, bluebells, foxgloves and nightingales are the well-known casualties, but populations of butterflies are also crashing. I'm afraid I wouldn't know a heath fritillary if it attacked me in a library, but it's dependent on our coppiced groves for its future survival. I suppose you could say that with so many others from almost every habitat under imminent threat of extinction – the latest are the once-common wall butterfly and the grayling – it's hard to know where to look or what to do. Well, we could start by only consuming what we need, but that seems to be a tall order. Predictably, the changing climate is also to blame: freshly energized grasses, brambles and bracken are swamping the coppiced ground and throwing everything out of balance.

Perhaps Knockwood is reverting to the original British wildwood. It will take several centuries, and the spiralling chestnut trees – with a thick undergrowth of sycamore – are not native to Britain, but this is where it's headed. Several trees have fallen and have not been cleared. There are bright-green ferns growing wetly on the decaying trunks. Where the tree still lives, new branches are bustling skywards. A silver birch has toppled and its exposed roots are dripping with Kentish clay. The good earth of home! It seems strange to think, but before the Great Storm of 1987, most people believed that a tree's roots spread as far and as deep underground as its trunk and branches filled the air. The tap root – the first one that plunges straight down from the germinating seed – supposedly mirrored the trunk above ground. The rest of the roots, a filigreed network, were thought to mirror the branches and twigs. If you could somehow fold over a winter tree at its base, and bend it underground, the spread of its branches would exactly match its own roots. I am sure I was taught that at school, but it's extraordinary that no naturalist shouted loud enough to be heard: 'Don't be ridiculous! Just dig down, or look at the roots of a fallen tree. It's not true!' The mighty Roger Deakin, who wrote his magnificent paean to the woods *Wildwood: A Journey through Trees* in 2007, pictured unimaginably vast, rooty depths, and wondered – as he slept one night with his ear to the ground – how deep they might go.

Well, it turns out that the roots don't go so deep after all. In fact the roots of some trees – the beech, for example – will snake along close to the surface, slithering over obstacles, and avoiding the rocky depths altogether; in some cases it's almost like the tree is *standing* on its roots. The root systems of trees are much smaller than we once thought, but they are vastly expanded by the mycorrhiza fungi that follow, connect and sustain them. It is an inspiring thought that trees have evolved to work with these fungi – they bring the trees nutrients from the soil, and the trees reciprocate with food from the sun – and neither can flourish, or even survive, without the other. Oliver Rackham suggests that some tree species would have colonized Britain faster, only they had to wait for their mycorrhiza to catch up. Every species has its own. It is almost certainly some flaw with the native oak's mycorrhiza that means its acorns can no longer germinate in the woods.

The old coppiced trees in Knockwood are growing fast. There's a hornbeam here that can't have been touched for over fifty years and there are at least six formidable, hard trunks growing from the same base. There is a line of hazel trees fluttering their coarse, late-summer leaves at the edge of the path that leads to the fields. The cob nuts, their smooth green heads rising pertly from the surrounding sheath of green leaves, will soon be ready for the squirrels, but not yet – and anyway, I quite like them at this time of year, once you get past the bitter pith. The hazel is apparently the only native nut that can be relied upon to ripen every year ('to swell the gourd, and plump the hazel shells/with a sweet kernel' wrote Keats). It's certainly true that the sweet chestnuts that fall here in such profusion are only occasionally large enough for roasting and must have been a sad disappointment to the Romans who introduced them. But what did they know of our English summers?

Hazel trees, though, are reduced to growing on the fringes of woods – or living on as coppices. They cannot compete with the huge shade thrown by ash, oak and beech. The hazel is one of the old sacred trees, worshipped at one time as the tree of wisdom, its nut proclaimed the food of knowledge, its forked sticks used for divining water and treasure. There's a lively energy to hazels – they were sacred to Hermes, the quicksilver messenger of the Greek gods – and they're an easy tree to sit underneath on a late-summer day, watching the sheep in their Kentish fields. Mr Longley used to protect his vegetable beds from the birds with string and strips of tin foil tied to hazel rods (I wonder whether there was some lost folk magic that made him choose hazel over any other tree, or was it just because they were whip-straight?), while my mother set her sweet peas clambering up tall hazel poles. H.L. Edlin, though, had other ideas: 'the hazel gives no quarter and it is simpler and often cheaper to cut it all flat and replant the ground with a smother-crop of some fast growing conifer such as Japanese larch or Douglas fir'. We lost a lot of hazel in the past century, but not, I'm happy to say, by my hand. That vandal, Wordsworth, is another matter:

> Then up I rose,
> And dragged to earth both branch and bough, with crash
> And merciless ravage: and the shady nook
> Of hazels, and the green and mossy bower,
> Deformed and sullied, patiently gave up
> Their quiet being: and, unless I now
> Confound my present feelings with the past;
> Ere from the mutilated bower I turned
> Exulting, rich beyond the wealth of kings,
> I felt a sense of pain when I beheld
> The silent trees, and saw the intruding sky –
> Then, dearest Maiden, move along these shades
> In gentleness of heart; with gentle hand
> Touch – for there is a spirit in the woods.

FROM 'NUTTING'

As Wordsworth says, it's too easy to confuse what we think now with what we might once have thought, but there's no doubt that the seeds of the present are planted deep in the past. The 'child is father of the man' – and whatever our age we all know there's a spirit in the woods, for good or ill.

My brother and I made a map of this wood. We'd come here on our bicycles and we drew (or I guess my brother did) an intricate chart of every path, hill, ancient tree and landmark. One time we came with my aunt Biddy, who must have been over fifty at the time but was game for anything. I remember her launching her bike off the edge of the steepest slope in the wood, the back wheel slewing wildly through the mud, before crashing into a clump of whippy coppicing at the bottom, breaking or spraining her wrist. She came back a month later, on foot this time, to show us what she'd done, and while gesturing down the slope she lost her balance and slid back flailing into the same coppice. She hurt herself again, although she tried to pretend otherwise. I can see now, as I peer down at the place, that it must have been a stand of hazel.

There are ghosts all over this 'roaring wood of dreams', as Housman once wrote. Many of the things I remember about my father seem to have happened or emerged while we were walking in the fields and woods; it is easier to share your feelings when you can walk and look away and examine the horizon; and we were right here when he told me – it was the kind of thing he said – that he didn't suppose he'd ever come to Knockwood again, he was getting too old. Who knows if this was true. His legs were certainly weakening. But I feel safe and at home here and when I look up there is sunlight in the treetops. Is he here? It would be just like him not to make his presence felt.

There was one other wood that I visited all the time when I lived in Tenterden. This was the walk I did most often. To start, head down Sandy Lane, the private road that leads past the balsam poplars (with their sweet summer scents), keeping the playground and tennis courts on your right (not to mention the 'new' 1980s leisure centre), and the houses and then the orchards on your left. If you peer through the orchards you should get a glimpse of the manor house, possibly Tudor, where my brother had a summer job picking apples for an angry German farm manager called Heinz and where the kind owners used to let me play (badly) on their tennis court, up until the day I leaned over the net and broke the rusted net support. Sorry about that, Mr Robson, it *was* me. Walking on, you will see that you are still on a road, but it is now rough and broken, and the hedges on either side are thick and high with dog rose, goose grass, hawthorn, nettles, ash saplings and ivy. There are some mature oak trees to keep you company. Walk straight ahead, and then, when you get to what was once a crumbling, Tolkienesque Kentish house, sunk deep in its grounds and gardens (it's now a rather more dapper affair), turn down a narrow, high-hedged path, making sure that the orchards are still on your left. There are usually sheep cropping under the apple trees' stooped, lichen-hung boughs – you are now far enough from home and safe to light your first teenage cigarette of the walk. On the right are fields and more sheep; the muddy path, after a few turns, will take you to a small stream. Do not step across the wooden plank and into the ('new', 1980s) housing estate, but instead turn right through a lovely little strip of steep woodland. There are hornbeam here, once but no longer coppiced, and many holly trees. There is fragrant honeysuckle twining in the fences up the slope to your right. At the end of this short woodland strip you must turn left and walk through a pear orchard. You are bearing slightly to the right. At one time, the pears grew thick and juicy in the trees, but today they look old and neglected, or maybe even poisoned, and there are just a few diseased fruit turning brown in the branches. I suspect developers are to blame, but no matter. Keep right and ahead and you will pass into the remnants of another orchard before reaching a farm track. Turn right and immediately on your left is the steep opening to a wood. Brush aside the feathery curtain of hawthorn and plunge down – and in.

My secret wood is not so secret now. At this end of the wood, the paths are well trodden and wider than I remember and there's a rope hanging from a sycamore tree, with a crisp wrapper stuffed into a hole in its trunk. But it always felt like an adventure here – and it still does. Woods are topsy-turvy places, where strange things can happen and the normal rules of human behaviour can be suspended for a while. When Alice goes through the looking glass she arrives in a wood and finds that everything – even herself – has forgotten its name. The lines between the species have dissolved. She hugs the soft neck of a fawn and they wander together under the trees (or whatever it is they're called) until they come to an open field, where the fawn cries out in delight: 'I'm a Fawn!... And, dear me! you're a human child!' and it gallops away. It is safer, but weirder, in the woods.

A small stream drifts through these woods and in the springtime its banks are busy with bluebells, wild garlic and the clear yellow stars of celandine. In late summer, though, the trees and the woods are calmer, the leaves and the forest floor a deeper green, the flowers ripened into seeds or fruit. The woods smell darker, and the shade is more profound, although there's a lingering twist of garlic in the air and perhaps the dusty scent of dry ripe wheat from the fields up above. The earth is still busy, but its job is almost done for another year.

There are rabbit holes everywhere, but no White Rabbit. My dog (a squirrel obsessive) is blundering about wildly, more excited than I've seen her for months. The path, as we move deeper into the wood, following the stream, is wavering and narrowing. There are branches to clamber over and there is thick, claggy Kentish clay to wade through. I have a memory of a lake and a heron – I know this is where the stream leads – and I am chasing it. A fallen tree – another sycamore – has hurled itself across the path. There are thick branches and forests of leaves to crawl through. I find I have lost my notebook, which I'd filled earlier with memories, and I have to turn back to look for it. I wonder, fleetingly, if the wood is reclaiming it – or them. There are more fallen trees blocking the path – a vast oak, roped and corded with massive effusions of ivy – and the path seems crazily hard to follow. It is damp. In fact, autumn seems to have arrived in the past hour. The ground is suddenly slick with rotting leaves and there are plates of fungi sprouting from every trunk. The lake is so much farther than I remember, although I am older – and slower – now. There's no avoiding the fact that I have aged much faster than these trees.

The path wends its slim and slippery way past the trunks of once-coppiced sweet chestnuts. There are bright red, white-spotted toadstools and layers of wet moss at the foot of the trees and I am *thrilled* to find a tiny oak sapling growing eagerly in the mud. Is there anything in this world so achingly hopeful as the slender thread of a young oak presenting five fresh green leaves to the wavering sunlight on the woodland floor? So much for Rackham and his mycorrhiza. A lime tree has fallen on top of a new wooden bridge, marked with yellow footpath signs, which someone has built very recently over the stream. My secret wood has been opened up, but the wood seems to have its own ideas. There's such vivid, exuberant growth here, it looks more like a jungle than a Kentish copse.

My father is still circling. There was only one wartime story he chose to tell us and it took place in the jungle. He'd been posted to India in 1941, and then on to Burma, commissioned as a second lieutenant in the Royal Artillery. He'd written in the flyleaf of his copy of *Indian Armed Forces in World War II: The Retreat from Burma 1941–42* that they had been 'trained (if at all) for the Western desert'. As they retreated from the advancing Japanese, still fighting but desperately low on supplies, they all contracted early stages of scurvy. The disease worsened by the day and they grew weaker until, miraculously, deep in the jungle, they came across a lime tree (the tropical kind) laden with ripe fruit. They gorged themselves and the scurvy vanished; and the next day they found they had the strength to escape back into India.

I have no idea as to the accuracy of this story. In retrospect, it sounds like something my father might have invented or embellished to tell his young sons, just so he could safely answer our occasional questions. One thing is for sure, he was happy to be invalided out and set to work in the Bombay docks. The only time he did try to share his memories was about a year before he died, when he was alone with my aunt Biddy and he suddenly, unexpectedly, started talking about the war, not stopping until late into the night. Sadly, though, he was an inveterate mumbler and her hearing was shaky, so we can only imagine what he might have been trying to say. Remembering this now, I realize I don't mind; but I do often think about him when I see a lime fruit.

I am still following the stream, looking for the lake, but we're not there yet. I am wondering, as I walk: how would any of us describe our ideal wood? Mine, I think, is an early summer wood, perhaps late May, when the leaves are out and fresh and there are flowers on the brambles and butterflies in the glades. The ground is dry on the path and there are spilt cushions of pine needles where I walk. There are oak trees, of course, but also lime, ash, hawthorn in the clearings, stands of hazel, red squirrels in a Scots pine, a bee humming drowsily in a clump of improbably early foxgloves. There has to be honeysuckle, and violets, bluebells shimmering up the hills, paths leading somewhere and nowhere, badger setts, perhaps a distant glimpse of a boar and its young, tufts of thistledown drifting in the slanted sunlight. Some of the trees are huge, the oaks in particular, masters of the forest, with heavy furled roots and branches you can swing and sleep on. There's a river, surely; and cool pools where the kingfishers dip and the otters splash. There are beavers! Birdsong fills the wood. I would like to know, I really would, that somewhere far away in this wood there are wolves. And eagles circling high above the trees. This wood never ends, of course, but stretches away and away. Are there people in the wood? Who makes the paths? I think we need people, perhaps some charcoal burners, a hut in the clearing, indeed a small village and (I'm thirsty now) a tavern. The woods are no good without people – if you ask me – although people have done very little good for the woods.

It's a childish dream of a fairy-tale wood. I may be back in Narnia, or I may be dredging up memories of my earliest walks, when I scrambled over the fallen trees and scraped out fishbones from the sweet chestnut leaves. But, I am bothered by Edward Thomas's idea that we should mistrust these visions. I don't know how much of my love of the woods is tangled up with a half-remembered, idealized childhood; and I have no idea whether there ever was a golden age when the woods bloomed in fragile harmony with humanity (although it's possible); but I do know, in fact I am absolutely sure, that there has never been a worse time for the woods than now. Take a look around. They have dwindled to almost nothing. Those that are left are isolated and threatened. The ash, larch and the oak trees are dying. The weather is convulsing and the new climate (whatever it becomes) is going to flood, scorch and starve countless species. Millions of conifers still suffocate the soil, the wildlife is being eradicated, noxious waste leaches and burns from the farms and cities, strange new diseases and parasites infest the plants. We are oblivious, or uncaring. All of this has been said already, but we still don't know when or how to help – or to stop. In fact, we don't want to stop. Who wants to say 'Enough!' or 'Too much!' – and that there are now too many roads, houses, businesses, shops, clothes, trucks, factories, books, guns, pens, pets, TVs and crisp bags? Who wants to say there are too many people? Too many people living the lives we live… The juggernaut of human activity is rolling and it seems unstoppable. There *was* a time when things were better. There was. If not for us, then for everything else other than the carpet mites and the sheep.

The deeply unfashionable historian G.M. Trevelyan, an early supporter of the National Trust, was convinced that there had been a golden time. (He came from a family of nature lovers – his brother Charles, the Lord Lieutenant of Northumberland, would stride naked around the Northumbrian moors.) 'What a place it must have been, that virgin woodland wilderness' of Anglo-Saxon England, he wrote in his 1926 *History of England*, 'still harbouring God's plenty of all manner of beautiful birds and beasts, and still rioting in the vast wealth of trees and flowers – treasures which modern man, careless of his best inheritance, has abolished, and is still abolishing, as fast as new tools and methods of destruction can be invented'.

I resent the fact that these thoughts intrude into the woods. At some point in our story, we stopped living with the trees, plants, animals and planets; we even stopped fearing or revering them. We just grabbed what we could and pulled up the drawbridge; and now we urgently need to find our way back. If, that is, they'll have us…

I arrive at a field of ponies and then, suddenly, at the lake. It is exactly as I remember. My father (him again) once bought a copy of Ivan Turgenev's *Sketches from a Hunter's Album* to take on holiday, which was so far removed from his usual reading diet of P.G. Wodehouse and Winston Churchill that we all wondered if he was having a late-life change. He was sixty-six years old ('Michael Fiennes 31/7/78' he has written in the front). The lake feels Russian. Grey and wild and empty. I did come here with him once. I must have done. A great grey heron lifts itself from the far bank and disappears into the distant woods. Coots scatter for safety. There are young oak trees by the path and thickets of reeds swamping the little stream. Bullrushes lean sturdily on the banks. Alder, their roots submerged, grip the muddy fringes and trail soft green leaves in the grey waters. There are water skaters slipping and darting on the surface. This is a sacred place, a long way from the nagging clatter of the world.

At the far edge of the lake there is a bank or dam, but I never liked to finish my walks by going there. There's nothing wrong with it: just slightly neglected fields, and woods, and a thickly rutted farm track, but I think I was always hoping for something better. Perhaps beavers. That would be the right ending to this walk. The dam would be so busy and well cared for and tended. And I am wondering, do beavers foul their own nests? Do they knock down their own houses, obliterate their own woods, poison their own land? And if they don't, what does that say about us? Put it this way: if you live in a place, are you more likely to cherish it? How close to somewhere do you have to live before you feel inclined to look after it? Is it your own home? The next street? The local neighbourhood? Is it the closest field or valley? Is it a county? A country? A planet? Or do you just take what you want and move on?

5

God gives all men all earth to love,
But since man's heart is small,
Ordains for each one spot shall prove
Belovèd over all.

RUDYARD KIPLING,
FROM 'SUSSEX'

There'll Always Be an England

It is a mysterious truth (and one the experts struggle to explain) that Britain has fewer woods than just about any other European country – and yet it has vastly more ancient trees. We have destroyed our forests, while opting to spare a large number of individual trees, most frequently the oaks and the yews. Perhaps the two facts are connected. Over the centuries, there must have been thousands of moments when a woodcutter was eyeing up an ancient tree, axe in hand, and yet decided – once again – to leave it alone, while our French, German or Italian counterparts just started swinging. And this is why, when you walk in the bountiful broadleaf woods of Normandy or Lombardy, you will find you are looking at trees of an invariably uniform age and size – it's rather like walking in those damned conifer plantations whereas, lurking in the last few woods of Britain, and not just in the woods, but in the fields, hedgerows and car parks, there are giants. This is something we have chosen to do. It can't just be because we harbour trace memories of the time when trees were sacred. That is important, for sure, but it's also true of the rest of Europe. Perhaps what is more significant is the British history of common land. Why would you want to destroy something of value and beauty that belongs to you? The ancient trees all show signs of having been lopped, pollarded and harvested: but they were not felled. So even when the land was enclosed, or some absentee landlord was vandalizing his own woods, it's possible that the local people tried to save what they could. We can but dream.

There must be spiritual reasons. John Stewart Collis, writing in *Down to Earth*, thought that the love of nature was embedded so

deep in Britain because we sense there's something hidden, waiting
to be revealed. This is true. We gaze at trees, we hug them, we lean
our backs on their welcoming trunks, we lie on the ground above a
busy mesh of roots, staring at the lulling branches – we are enfolded
by trees. Perhaps we are waiting for them to answer our questions.
But can we really say it is an exclusively British trait?

Even so, there are something like two thousand ancient yew
trees growing in Britain, while there are only about one hundred left in the whole of the rest of Europe. Some of them are
thousands of years old. And oaks! There are 112 'great oaks' in
England (the 'great' means that the trees are over eight hundred
years old); but there are only about eighty of them to be found in
a vast swathe of land stretching from Calais to Cadiz and Athens.
What's more, we are uncovering ancient trees in Britain all the
time. Where have they all been hiding? Do they think it's now safe
to come out? Of course, we shouldn't get too carried away. A little
over one hundred great oaks left in the whole of England – that's
compelling evidence of a massacre. But it is nonetheless something to celebrate.

I am standing in Windsor Great Park, contemplating the oak
trees. The setting is typical wood pasture: large, well-spaced, venerable pollarded oak trees; fenced copses of mixed birch, beech
and hazel; and open stretches of deer-cropped grassland. There is,
I realize now, another possibility: in Britain, do we love the trees,
but fear the woods? Have we cleared the undergrowth and thinned
the trees, so we can stand back and admire a few mighty specimens
free from the hidden menace of feral outlaws and boggarts? Is
this what we're really after? Pasture and parkland and a republic
of sheep? It's possible, despite the enthusiasm of 'rewilding' dev-
otees – those of us who would like to see the reversion of Britain
to a more primordial state, bristling with elk, boar, wolf, lynx and

bison – it's possible that what the majority really want is more *order*; and perhaps what we'd all secretly prefer is something along the lines of a Sunday night BBC country estate, complete with grand trees, sweeping lawns, a table set for tea, a chewing of cows just beyond the ha-ha and Tom Hiddleston's sumptuous white buttocks humping rhythmically in the shrubbery.

Whatever the reason for the survival of so many ancient trees in this country, it turns out that most of them are not protected by the law. There has been some discussion about offering 'owners' financial support if they promise not to dynamite the oak tree at the end of their driveway, but the fact that cash benefits are being discussed shows the real threat to these extraordinary trees. We will only protect what we value – and we have forgotten that 'value' is not a financial term. The law is essentially irrelevant if we can't agree what actually matters. If you can imagine a society led by Druids – and I mean real Druids, not the ones who work in an office in Hemel Hempstead before donning white robes and a ram's head at the weekend – such a society would not fell their ancient trees. If our trees are to survive far into the future, we all need to spend more time in the woods. We need our children to climb trees, swing from branches, light fires, carve their names in the bark and kick fir cones through the glades. Even one single visit can be transformative and imbue feelings of belonging and ownership. We should think of taking our children to the woods as an investment, in the nicest possible sense, in the future. This, of course, is incompatible with our schools' narrowing National Curriculum. And it's also incompatible with a study that showed that over the course of four generations the children in a typical UK family have gone from walking eight miles per day to never leaving their back garden (if they have one). It's no wonder that over eighty per cent of UK schoolchildren can't identify an ash tree – or that so many of them think of parks, woods and commons as dirty, frightening and unpleasant places.

That said, I am sure that most of us would have struggled to identify an ash tree when we were at school, and it seems unfair to cluck in horror at our over-tested and much put-upon children for the same reason. I am certain that they'd have much better luck with an oak – and where better to start than with the Windsor Great Park oak trees, supreme examples of their kind? There are about five hundred species of oak living in the temperate forests of the northern hemisphere (plus a few hybrids), but there are only two that are native to Britain: the common or pedunculate oak (*Quercus robur*, also known as *Q. pedunculata*) and the sessile or durmast oak (*Quercus petraea* or *Q. sessiliflora*). Incredibly, about one-third of all the trees living in the woods of Britain are oak, although there was a desperate lull in their planting during the Locust Years. If you really feel the need to tell the difference between the two natives, the pedunculate oak is happier in south-ern and eastern Britain, and prefers to sink its roots into deep, heavy soils (even clay). If you're looking at one when the acorns are growing you'll see that they are suspended at the end of long stalks; at other times of year, you should look for their hairless twigs and the deeply lobed leaves carried on short stalks. The sessile oak is happier in the north and west, in thinner (and more acidic) soils. Its leaves are larger, the lobes are shallower and the stalks are longer (although the acorns have no stalks). Its twigs are relatively hairy. Here's a mantra to help you remember: deep, southern and hairless; shallow, northern and hairy – although, to confuse everyone but the most dedicated naturalist, the two often breed and create hybrids.

Famously, a mature oak tree can support over three hundred species – I have read it's even as many as five hundred – and if you consider that we use them for beams, boats, doors, windows, floorboards, barrels, fencing, gates, chairs, tables and beds (in fact, most furniture), the props and chocks for mining, bowls, casks, tubs, moulds, spoons, panelling, charcoal, firewood (deadwood could once be removed from the lord's local wood 'by hook and crook'), tanning (from the bark), food (for us in times of famine, and for our pigs at all other times), medicine (it is said that walking around a large tree muttering incantations can cure rheumatism), worship, hiding (for commoners and kings), handles, mallet heads, wedges, roofing, poles, spokes, the steps of ladders, fine turnery and wood carving, church pews and roofs and the iron-hard sides of railway wagons and warships – if you consider all that, then you can include humans in the list of species that find the oak tree indispensable. The oak forests haunt our imaginations; and we have named countless towns, villages and landmarks after this preposterously charismatic tree. The Celts and the Welsh called the oak 'darach' and 'derw' – hence Druid and Derry – while the Saxons used 'ac'. There are forests of oak buried for ever under London, from Acton in the west to Hackney in the east, while people gathered to worship, preach and picnic under Gospel Oak in north London and Honor Oak in the south.

It is said that an oak tree takes three hundred years to grow, three hundred to stand still and three hundred years to die. In that time it will be grazed by deer, cattle and grey squirrels, and beset by countless invertebrates, especially moths. There's the acorn and the winter moth, as well as weevils, butterflies (such as the oak specialist, the purple hairstreak *Quercusia quercus*), bees, beetles (including the oak-dependent click and chafer beetles), wasps (with their galls), aphids, numerous fungi (including the beefsteak fungus), birds such as the great spotted woodpecker feeding on the insects and nesting in the branches, badgers and boar snuffling in the roots, jays carrying the acorns far and wide, and the voracious oak tortrix (*Tortrix viridana*), a moth that is also known as the 'green oak roller' and can swarm all over the tree in the summer, even though in its turn it is hunted and consumed by titmice and the ichneumon wasp. Other hungry caterpillars include the mottled umber with the descriptive Latin name *Errannis defoliaria*. An oak can be so beset by larvae that it loses much of its foliage in midsummer and is forced to put out a whole new set of leaves later in the year. Anyone sitting under an oak tree at this time will find themselves generously bathed in a soft, pattering rain of caterpillar crap. It is a bountiful tree, and mostly untroubled by these attacks, so long as they do not happen too many years in a row. The new, compensatory leaves and shoots that spring from the oak after they have been devoured are known as Lammas growth and they coincide with the Celtic harvest festival of early August.

But beyond all that, and to the glory of us all, the oak is the foundation of the Royal Navy and the cornerstone of the nation. It is the impenetrable oak wall behind which lurk men of unfeasible courage, sustained by their hearts of oak. It has been our bulwark and our national metaphor for longer than we can remember. Its immense, iconic power cannot be co-opted or diminished by the brand-obsessed spivs of the present-day Conservative Party. The Old English rune poem says it well:

> The oak nourishes meat on the land
> for the children of men; often it travels
> over the gannet's bath – the stormy sea tests
> whether the oak keeps faith nobly.

There are also these more famous (but surely rather clumsy) lines from Tennyson:

> There is no land like England,
> Where'er the light of day be,
> There are no hearts like English hearts,
> Such hearts of oak as they be...
> And these will strike for England
> And man and maid be free
> To foil and spoil the tyrant
> Beneath the greenwood tree.

FROM *THE FORESTERS*

The oak tree and the English 'man and maid' stand stubborn and free, united against the (foreign) tyrant. Britons won't be slaves. And the national poets continue to tie themselves in knots over whether they are extolling 'Britain' or 'England'. Here, at least, is a British oak –

'It is a striking but well-known fact,' wrote John Charnock in his three-volume *An History of Marine Architecture* from 1800, that:

> The oak of other countries, though lying under precisely the same latitude with Britain, has been invariably found less serviceable than that of the latter, as though Nature herself, were it possible to indulge so romantic an idea [oh, go on, please do], had forbad that the national character of a British ship should be suffered to undergo a species of degradation by being built of material not indigenous to it.

Throughout the Napoleonic Wars, when Britain's hunger for good-quality timber was at its height, Nelson's navy had to import large quantities of Swedish oak. The navy also relied on capturing French, Spanish and American ships, which seemed to serve them perfectly well. But the myths grew and became embedded: British (English)

oak is best; and because of this undoubted superiority we lost many
of our oak forests to the ravenous shipyards. The second myth
seems easier to justify. After all, a single seventy-four-gun warship
consumed about two thousand mature oak trees in its construction.
The oak tree was especially useful, not just because of the hardness
of its timber, behind whose wooden walls the hearts of oak sheltered
and roared, but because the angular and forked branches were
indispensable for almost every part of the ship. Oliver Rackham
has proved to his own satisfaction that the demands of the navy had
no significant effect on the loss of Britain's woods, but there was
without doubt a great slaughter of *trees*. What is also certain is that
along with our love of the oak tree there has for centuries been a
parallel anxiety about its disappearance and decline. John Evelyn's
Sylva, or a Discourse of Forest Trees was written in 1662 for precisely that
reason: to urge people to plant more oak trees. Not only were they
in peril, but they were essential for the nation's future prosperity and
health. He creates a lovely image, urging us all to make sure that:

> His Majesty's forests and chases be stored with this spread-
> ing tree at handsome intervals, by which grazing might be
> improved for the feeding of deer and cattle under them,
> benignly visited with the gleams of the sun, and adorned
> with the distant landscapes appearing through the glades
> and frequent valleys; nothing could be more ravishing. We
> might also sprinkle fruit-trees amongst them for cider.

No one paid much attention to this fruity panegyric, although in
1669 (when Evelyn brought out the second edition) he himself
claimed that he had been responsible for inspiring the planting
of two million trees since his book was first published. This was
ridiculous – Evelyn halved the number a couple of years later – but

even if it had been true, the real crisis in Britain's woods continued to unfold. A survey taken in the year 1608 in some of Britain's most important remaining forests, most of them helpfully placed for the shipyards – the New Forest, Sherwood Forest, Alice Holt, Forest of Bere, Whittlewood Forest and Salcey Forest – found exactly 232,011 mature oak trees; by the year 1783 (and that's before the final urgent expansion of the navy, let alone the full impending onslaught of the Industrial Revolution) there were only 51,500 oaks left. The oak was favoured by foresters and retired admirals, and idolized by patriots; but in the end it was found to be far better laid flat on its back and stripped for timber, on its way to making a nice fat profit for its owner. I'm with John Evelyn: anyone who destroyed 'those goodly forests, woods and trees' was to be condemned to 'their proper scorpions… to the vengeance of the druids'.

The oaks in Great Windsor Park have small metal number tags nailed to their trunks. I find myself in front of 'Number 00034', standing at its base, stroking the rough bark, marvelling at the sheer looming strength, the overpowering energy and size of this earth giant. If you stare with eyes half-closed, can its thick, rippling bark transport you to the storm-tossed Atlantic seas? Can you smell the salt spray? There is a connection between the oak and the ocean – we have set sail many millions of times over 'the gannet's bath' to test whether 'the oak keeps faith nobly'. But the connection runs deeper than that, and humans are peripheral – irrelevant – to the relationship. In 1998 the Japanese marine chemist Katsuhiko Matsunaga laid bare the planet-defining partnership in an essay called 'When Forests Disappear the Sea Dies', here explained by Jim Robbins in his book *The Man Who Plants Trees*:

> Phytoplankton… are vital to all life on the planet. They are
> abundant in the world's oceans and, through their process

of photosynthesis, they are tasked with turning sunlight into food for other sea-dwelling creatures. For this, they need iron – but even where iron is abundant in parts of the ocean, it is oxygenated, which means it is not readily available for the tiny creatures. What *can* make iron available to phytoplankton to perform photosynthesis, however, is fulvic acid, one of several humic acids that comes from the decay of leaves and other organic matter. The ongoing, natural decomposition of centuries of tree leaves and other material on the forest floor, and the leaking, leaching, and washing of this chemical stew into the ocean, is vital to increasing coastal phytoplankton, and thus the things that eat them, and those that eat them, from oysters all the way to whales.

Ever since Matsunaga's revelations, Japanese fishermen have been planting trees along their coastline and marine life is returning to the oceans – although the whales may want to consider whether there aren't more secure places to live than in the seas next to the world's most lethal 'whale research' fleet. It's just a shame that this kind of world-changing information – the fact that we will all die if we persist in destroying our forests – very rarely reaches most of us. And if it does it's presented at the end of the news, as some kind of joke, jostling for space with the latest bulletins from the worlds of fashion and football. I'm more likely to hear about some celebrity's new tattoo than I am about the end of the forests.

Incidentally, is David Beckham turning into a tree? In Ovid's *Metamorphoses* there were plenty of young people (women, mainly, but still…) who were transformed into trees as they fled from the priapic gods, or from their own crimes, their skin spreading with filigree patterns, changing colour and coarsening:

For even as she prayed, the earth closed over her legs; roots grew out and, stretching forth obliquely from her nails, gave strong support to her up-growing trunk; her bones got harder, and her marrow still unchanged, kept to the centre, as her blood was changed to sap, as her outstretching arms became long branches and her fingers twigs and as her soft skin hardened into bark.

'MYRRHA TRANSFORMED TO A TREE',
FROM *METAMORPHOSES*, BOOK X

In the world of Ovid, people are closer to the natural world, to the animals and trees, the rivers and their gods. The boundaries are permeable. Perhaps today, as we strip away the last of the world's wilderness, many people are trying to re-create it on their skins.

Back here in Windsor, it is easy to see that the oak has an outer layer that was able to withstand the heaviest blows from the hungriest super-elephants that roamed the Great Park one hundred thousand years ago. It evolved to evade these and other predators, branching and coppicing if eaten when young, putting on great slabs of ridged armour in older age. We've yet to see what it will do about the chainsaw and the diesel car. This oak, 'Number 00034', has tufts of bracken sprouting from its bark like the plumes of knights' helmets and, looking closer, it's possible to see where woodworm are drilling their small holes in patches near the base. But nothing is going to slow or topple this tree for another hundred years or more, not even the lightning that must have struck it hard in the past, shearing off its main trunk. It's a battered tree, with just one mangled stump for a bough, its innards exposed and bleached like Japanese whalebones, but it is vigorously alive.

The (spiritually) thirsty poet William Cowper compared his hollow 'Yardley Oak' to 'an huge throat, calling to the clouds for drink':

Time made thee what thou wast, king of the woods.
And Time hath made thee what thou art, a cave
For owls to roost in…
But the axe spared thee; in those thriftier days
Oaks fell not, hewn by thousands, to supply
The bottomless demands of contest waged
For senatorial honours. Thus to Time
The task was left to whittle thee away
With his sly scythe, whose ever-nibbling edge,
Noiseless, an atom and an atom more
Disjoining from the rest, has unobserved
Achieved a labour, which had far and wide,
(By man performed) made all the forest ring.

FROM 'YARDLEY OAK'

Close by is tree 'Number 00036', another hollow tree, but less generously so. The opening is narrower – and peering in I can see a grubby collection of old cans, strips of pipe and a couple of bulging dog bags, oozily stuffed with excrement. They are not exactly votive offerings. It seems like a violation, not least because many of the trees here (especially for some reason the younger ones) are twisted around with yellow ribbons and hung with memorial flowers, soft toys and laminated pictures and poems. Husbands, wives and children are all commemorated. Trees have always been planted or decorated to mark our passage through life and death, but Windsor is especially rich with memorials – I am wending my way towards John F. Kennedy's in Runnymede. Not far from here is a straggling, 140-year-old oak next to a fading metal plaque that reads: 'Planted by Prince Victor Christian Schleswig Holstein April 14th 1888 in commemoration of the coming of age of His Highness on that day.' Young 'Christle', as he was known to his family, would have been eleven years old on that day – a rather early 'coming of age', but they did things differently then. He was the eldest son of Princess Helena, the third daughter (and fifth child) of Queen Victoria, and (if you're still with me) he was the first member of the royal family to go to school. He grew up in Windsor Castle, so knew these woods and pastures well, and he turned out to be good at cricket, captaining his school and then his college at Oxford (mind you, who was going to suggest that he *shouldn't* be captain?). He fought in various wars and died of malaria aged thirty-three, in Pretoria, South Africa during the Second Boer War. Staring at his photo, there is no doubting he has the lugubrious, soft-boiled eyes of his family, but there is also a rare glimmer of humour there, I think. 'Windsor', of course, is now the name of Britain's royal family, changed from Saxe-Coburg-Gotha in 1917 at a time when the Germans were not exactly our nation's friends. I wonder whether the solid, oaken quality of 'Windsor', with its undeniably ancient, heartily *English* woods, was also what attracted this family of one-time German princelings to the name?

A short distance from this plaque, past lime trees draped with mistletoe, is another one, larger but just as faded. It reads:

THIS GROVE OF OAKS WAS
PLANTED ON 19 JUNE 1937 BY
BRITISH COMMONWEALTH REPRESENTATIVES
IN THE PRESENCE OF
KING GEORGE VI AND QUEEN ELIZABETH
TO COMMEMORATE THEIR MAJESTIES CORONATION
ON 12 MAY 1937

I had in fact reached this grove of oaks, all of them just over eighty years old, long before I found the memorial, and had been puzzling over the meaning of it all. It seemed like a very esoteric selection of trees, each of them underpinned by a plaque attaching them to an apparently random collection of (often arcane) places. The oaks had been planted in the run-up to war (and even then, eighteen months before Chamberlain's paper-waving, people must have known in their hearts what was coming), and it looks as though every country of the Old Empire had sent someone to plant an oak tree in an act of commemoration, but also, perhaps (is this fanciful?), in solidarity. Each nation, state or dependency has been allotted its own species of oak, marked with a metal plaque, although the species do sometimes overlap. Thus Trinidad and Tobago was given the red *Quercus coccinea* 'Splendens' (it has grown especially well), while the Seychelles was awarded *Quercus acutissima*. Some trees are doing better than others. The *Quercus palustris* of Perak (Malaysia) is frantically busy, although it is also rather a mess; while the tree that is thriving better than any other is the *Quercus sessiliflora* 'Muscoviesis' of Kedah (also now a state in Malaysia). Perlis (Malaysia's smallest state) gets the evergreen *Quercus ilex*; British Guiana has also been given the red *Quercus coccinea* 'Splendens', but poor old Kel's oak has been snapped in two. I am sorry to admit that I have no idea where Kel is (or if it even exists any more), but I am eager to match the trees to the states. What was the reasoning? Why did Bermuda, for example, get given the very English *Quercus sessiliflora* (although this particular specimen has a tropical stripe of bright-green moss growing up its front), while Malta got the *Quercus macrocarpa*? It is native to North America, after all. Does Malta not have its own oaks?

I also wonder if, somehow, the health of a nation affects the state of its designated tree. Or even vice versa. Do these trees need to be protected, like a fragile thread nursed by The Fates? I go looking for South Africa. Did it grow stunted and deformed through the years of Apartheid, before being transformed by a blooming miracle in 1990?

Sadly, its *Quercus libani* gives no clue. On the other hand, Penang's *Quercus palustris* (another North American native) is encircled by a blackened ring of dead earth. Are we witnessing a distant whisper of its intemperate industrialization? The matching of trees with states becomes obsessive. I see that the United Kingdom was awarded a *Quercus robur*, our southern oak, as was 'His Majesty the King' – who was deemed a nation for the occasion. I hope the north did not rise up in protest. I arrive at a section of troubled, disputed (or no longer existent) lands: Palestine, 'Transjordan', 'Aden', 'Burma', 'Tanganyika' and 'Northern Rhodesia' – they all have their oaks and their place in this grove. 'North Borneo' was given the hybrid *Quercus hispanica* 'Lucombeana', planted at a time when the place was a protectorate under something called the North Borneo Chartered Company, and five years before it was occupied by the Japanese. It is now part of Malaysia, but its tree lives on. I pass by Johor (and its attendant *Quercus palustris*), lamenting my ignorance. The final two states are The Straits Settlements and Sarawak (both twinned with the Western Asian *Quercus macranthera* – perhaps they ran out of species). These oaks are thriving – there is no sign of their native forest fires – but indeed all of the oaks in the grove look to be set fair for the next seven hundred years. They have outlasted every official who was there – surely – but also many of the protectorates, states and fledgling nations that they were planted to represent.

I have decided we read too much into trees. I blame Shakespeare, who loved a gardening metaphor more than most – especially one that could shed light on the rotten state of a nation. The lines from *Richard II* are some of the most famous (spoken by the 'First Servant', so you might assume they were meant to be played for laughs, but apparently not). Here he is grumbling that there's no point in doing any work on the garden because the whole country is going to rack and ruin:

Why should we in the compass of a pale
Keep law and form and due proportion,
Showing, as in a model, our firm estate,
When our sea-walled garden, the whole land,
Is full of weeds, her fairest flowers chok'd up,
Her fruit-trees all unprun'd, her hedges ruin'd,
Her knots disorder'd, and her wholesome herbs
Swarming with caterpillars?

SHAKESPEARE, FROM
ACT III, SCENE IV, *RICHARD II*

He should see it now. The tree lover and naturalist Gerald Wilkinson made his own version in 1973 in his book *Trees in the Wild*:

Our small country is no longer the waste land from which we grasp the necessities of life. It is like a garden – with a large plot reserved for food production, a few trees, a rockery, some open grass, some formidable sweeps of concrete, a copious rubbish tip and some odd corners where anything can happen.

This was written at the fag end of the Locust Years. Forty years on, we might want to add that the rubbish tip has now spread much further (and deeper), the concrete is abundant and formidable, the caterpillars are mostly dead, and that you wouldn't want to go swimming in the pond. But at least there are undeniably more trees, even if most of them are dainty little Christmas trees lined up neatly on death row.

These national metaphors grip us unconsciously hard. Britain is not the only country to imagine that its 'health' is somehow connected to the state of its woods, fields and farmland. Or that its geography is uniquely beautiful and blessed. *La Belle France* feels the same way, as do many others. It has often been pointed out that in the First and Second World Wars almost all of the iconography exhorting the nation to fight for its survival was pastoral. The government's posters showed the golden hills of a West Country harvest and the drowsy cottages of half-timbered villages, but almost never the factories and houses where most of the population worked and lived. There was, arguably, more justification for it then: many people had moved from the land in living memory (or their parents and grandparents had). Perhaps you could argue that the early twentieth-century Britons knew their country, its fields and its farms, much better than we know it today. The cities were smaller and huge numbers of people visited the country on foot, by bike and by train – urged on by the government and the railway companies and enthused by the newly formed National Trust and the Ramblers' Association. There was a mania for fresh air and exercise, from hop-picking working holidays in Kent to the scout camps of the Forest of Dean. 'It [the National Trust] belongs to you all and to every landless man, woman and child in England,' trumpeted its founder, Octavia Hill. A sense of ownership is all. And, in the end, people are more likely to march off to war if they are holding a vision of something that is theirs and worth saving. Only the Soviet Russians could imagine that people were going to be inspired to lay down their lives for an image of a belching chimney. Although they did, in their millions, just the same, for fear of something worse.

So, no, people are not inclined to foul their own nests. And that being true, we urgently need to learn to feel a connection to our land, its woods and its trees, if we are to have any chance of saving them. A wider, real, actual ownership might help too, but let's come back to that. Right now, what the woods need is for us

to rediscover a shared love for what we have; a benign local patri-
otism that brings us together, ready to fight for what's left. Naomi
Klein makes it abundantly clear in *This Changes Everything* (2014)
that it will be a hard and bitter fight: there is too much at stake for
those who are heavily invested in the continued exploitation of the
earth. 'A tree is a tree,' said Ronald Reagan, 'how many more do
you need to look at?' It is right to hold to account those who do
the most damage: the huge chemical-farming businesses, the fossil
fuel and mining companies, the easy-money developers. Without
exception, these people live far away from the scene of their activities. They live in Surrey and Manhattan, Singapore and Shanghai,
many comfortable miles from the dynamited hills of West Virginia
and the fracking front lines of Yorkshire and Balcombe. Even the
people who tip their toxic bin bags and their wrecked cars down
the woodbanks in lay-bys do not, surely, own that wood, or live next
to it, or walk through its groves every week. The people who are
fighting hardest for what is left of raw nature are the people who
live there and cannot or do not want to move: indigenous peoples,
small farmers, the families on the edge of the new landfill. Again, to
call these people NIMBYs is a PR sham. It's a pity most of us now
live in cities; the sights and sounds of the devastation are distant...
In Britain, we completed most of our deforestation centuries ago.
But we have surely all noticed that the list of natural places we can
visit is dwindling, at home and abroad. Klein calls the incalculably
huge, rapidly spreading lands we have lost to economic growth 'the
sacrifice zones'. So while it is right to focus on those who do the
most damage, the horrible fact is that we all know who that is. It's
us. The real question is this: is there anything we are prepared to
sacrifice to make it stop?

It would really help in this country if we knew what we were
called. There's a lot of power in a name. It is the constituent

parts – Scotland, England, Wales, Yorkshire, Roseacre Wood – that
are rolled out when our hearts are being tugged. And if there are
fights to be had, they will be local. It is very rarely Britain that is
evoked successfully. Scotland has Robert Burns and Walter Scott.
Wales has Thomas (Dylan) and Thomas (R.S.). Blake sings of
England and its green and pleasant land; and Kipling understands
the power of the idea when he quotes a soldier returning from war:

> If England was what England seems
> An' not the England of our dreams
> If she was putty, brass and paint
> 'Ow quick we'd drop 'er! But she ain't!

KIPLING, FROM 'THE RETURN'

It turns out that we need to sign up to a dream… but perhaps that
doesn't matter.

I find I have wandered to the base of what I think is tree 'Number
006536', although its metal tag is illegibly corroded. I am subconsciously admiring its mossy elephant's feet, the grooved, gnarled bark
(it's a sessile oak), its huge, incomprehensible heft, while I earnestly
turn over in my mind the need for the *whole of Britain* to unite in our
forests. It is a sunny day in November, but when I look up, and as
far as I can see in this whole sweep of gorgeous parkland, within
easy reach of London, there are almost no people. In fact, at this
moment there are just four middle-aged white men, with black
labradors, shuffling through the autumn leaves. Make that five,
including me, although my dog, while black, is wild and shaggy
and of indeterminate parentage. Of course, this is not fair. It's a
school day. And I have previously passed just as many women, not
all of them white or middle-aged. In any case, the fact that so many
people feel dislocated from the land is not a matter of gender or

age: it's one of poverty, ownership, class and, concomitantly, race. The county with the most National Trust members is Surrey. But it's still troubling.

I bring this thought to Judy Ling Wong, founder of the Black Environment Network (BEN), and she laughs at my worries. Of course, she says, if there is no sense of belonging then there is no care for the land and no stewardship. But this is not something that only affects ethnic minorities. It is poverty, not ethnicity, that predominantly affects our relationship with the place where we live – and disadvantaged groups are squeezed by a lack of time and money. Many people with low-paid, physically demanding jobs want to rest at the weekend (if they have one), not walk in the woods. Anyway, she says, everyone in Britain comes from a specific place, with its own culture and stories. The parts of the UK – she mentions Wales, Northumbria, Cornwall and Scotland – where the communities are stronger, they all have their stories and are more ready to defend what they have. And what does a tree care about the provenance or beliefs of any human being, so long as that human treats it with respect? Or, perhaps even better, leaves it alone.

'Nature stands on her own,' she tells me. 'Expose people to it and they will fight for it. We all love what we enjoy and protect what we love.' The love of nature can be awakened with just one visit, when someone sees a forest or the ocean for the first time. In fact, it is often people who have just arrived in the UK, who have come from a rural place or farmland, says Judy, who 'cannot wait to feel their bare feet on Mother Earth again. To be an ethnic minority is to be environmental.' In 1999 a local Sikh community planted three hundred oak trees in Khalsa Wood in Bestwood Country Park, outside Nottingham, to celebrate the three-hundredth anniversary of the founding of Khalsa, the worldwide fellowship of all initiated

Sikhs. 'They marked the place with memory,' said Judy, 'and that is what gives people a sense of belonging. Nature needs to be normalized and then you can kick-start the relationship. Displaced peoples planting trees is a poignant and meaningful claiming of their new homeland.' I am struck by the fact that the three hundred trees planted (and since then there have been more) were chosen by the Sikh community to be oaks, settling their roots in the English soil for what could be the next eight hundred years.

It's a hopeful thought, but we need a thousand more Khalsa Woods. 'When is the best time to plant a tree?' runs the Chinese proverb. 'Twenty years ago. The second best time? Today.'

Next to tree 'Number 006536' (or whatever it calls itself) is a small copse of silver birch, interspersed with large, flourishing clumps of rhododendron. The birch ('the ladies of the woods', as Coleridge called them) have grown tall and elegant, their last few autumn leaves flapping in a soft wind, so it is not quite as Coleridge describes the scene in 'Christabel':

> There is not wind enough to twirl
> The one red leaf, the last of its clan,
> That dances as often as dance it can,
> Hanging so light, and hanging so high,
> On the topmost twig that looks up at the sky.

Coleridge is describing an oak tree on a bitter cold night in April ('and Spring comes slowly up this way'); but I'm more interested in the rhododendron bushes that have grown so shaggily bold and are snapping at the trunks of the birch trees. The rhododendron has a dreadful reputation among woodland conservationists: an alien species that was usually planted in woods by gamekeepers to give ground cover to their soon-to-be-slaughtered birds, it has grown and spread rampantly, engulfing the native flora and suppressing the regeneration of native trees. I'm surprised the wardens of Windsor Park haven't dug it up, although it is notoriously hard to shift its tenacious roots. And it does look splendid in early summer.

Boethius was quite wrong when he wrote:

> Canst thou not see, that every herb and every tree will grow best in that land which best agrees with it, and it is natural and habitual to it; and where it perceives that it may soonest grow, and latest fall to decay?... Take therefore tree or herb, whichsoever thou wilt, from the place which is its native soil and country to grow in, and set it in a place unnatural to it; then will it not grow there, but will wither. For the nature of every land is, that it should nourish herbs suitable to it, and wood suitable.

> BOETHIUS (TRANS. KING ALFRED)

The British have an especially conflicted relationship with their introduced species. From Sir Walter Raleigh onwards, bearing his precious potatoes, no other nation has been so enthusiastic in hunting down new trees, plants and flowers, bringing them back home by the thousand to see if they will flourish and bloom. Britain's gardens are a multicultural delight. Our forests are an object lesson in exotic experimentation and the conifers, whatever their gruesome ecological impact, have been wildly successful in providing cheap, ready timber. But, still. Britain is an island – and we do like our wooden walls. We can list our native trees and we know what our land contained before the humans got busy. Quite rightly, there is a growing movement to restore our woodlands as close as possible to their original, post-Ice Age state and to nurture the native plants that are most likely to withstand the coming climate disaster. The free movement of plant and animal species has had unintended and disastrous implications for these efforts: so many distressing and miserable developments, from ash dieback to aggressively spreading rhododendrons, can be blamed on alien imports. In early 2016 we added the box tree caterpillar to our many anxieties. It is spreading in untold numbers from Asia. There have been sightings in Sweden! And now in south-east England! It can defoliate a box tree with its ravenous foreign jaws in just a few hours!! Close the borders!! In New Zealand, another island with a unique ecosystem, they swerve their cars on the roads to kill the introduced Aussie possums that are devouring the chicks and eggs of their native (and defenceless) birds. In North America the *Dendroctonus* bark beetle (literally, in Latin, 'tree killer') is flourishing in the warmer winters in places it never before lived and is laying waste to the forests.

So it's never entirely surprising when you come across a sign like this pinned to a tree when out walking in the woods:

SYCAMORE AND NORWAY MAPLE

AT ASHFORD HANGERS

NATIONAL NATURE RESERVE

These sycamore trees have been felled as part of an on-going management of the woodland.

Ashford Hangers is noted for its rich and diverse woodland flora including rare Orchid species. While these woodland habitats are robust in terms of adaptation to relatively short term change (woodland management, storm damage to trees, etc), it is considered that Sycamore *infestation* has the potential to permanently change the ecology of the woodland.

The italics are mine. The sycamore, you may recall, is not a native but was introduced to Britain in the sixteenth century. And because of our dangerous addiction to nature-based national metaphors, I find I am all of a sudden overwhelmed by an urge to make an embarrassingly obvious point. It's this. Trees are trees. People are people. And the box tree caterpillar is indeed the box tree caterpillar. Outside the poems of Ovid, there really should not be any need for confusion. But before I sink any deeper in a proliferating welter of metaphor, perhaps for now we can just agree on one simple thought: we should try not to make a woodland conservationist feel awkward when they are attempting to restore the ancient purity of the British woods; but, believe me, there are plenty of times when we do.

It is quite a relief to leave Windsor Great Park.

One of the reasons the sycamore is so successful in Britain is because it has no natural pests except for, somewhat ironically, the more recently introduced grey squirrel. We could keep on bringing in new species until we hit the right balance, although that approach doesn't have a happy history. Nature, like Nanny, knows best. On the way out of Windsor Great Park (I'm on my way to Runnymede) I pass a beech tree with a round and pouting burl on its trunk, puckered like a cat's anus. There's no more accurate description, but it makes me realize how much we like our nature imagery to be kept pure and Disney fresh. Possibly even frozen in time. Has a love of nature come to be seen as old-fashioned and esoteric? And not, perhaps, entirely essential to the real business of driving a modern country forwards? There was a predictable outcry in 2015 when the editors of the *Oxford Junior Dictionary* dropped the words 'conker' and 'acorn' from its pages, but I know from my own children that not only can they not recognize an ash tree, they think I'm 'weird' for thinking they might even want to. We have a long way to go.

Autumn leaves and rain fall from the trees as I trudge past the large, overindulged houses that fill the space between Windsor Great Park and Runnymede. They seem to have names like 'Winter Palace' and 'Meadow Grange', although it is hard to tell because my spectacles are smudged and clouded by the drifting rain. Rows of stately Scots pine have been arranged into mini-plantations, but the pine is an unpredictable and wayward tree and does not take kindly to straight lines. Holly and beech, both trees that are more easily controlled, have been fussed into tall, thick hedges. It is difficult to shake off the feeling that these trees – this whole neighbourhood – would have been better off left as woodland.

I am visiting Runnymede, quite by chance, on the anniversary of John F. Kennedy's assassination. His simple memorial stone is placed in a lovely patch of 'English Woodland', as the notice describes it, at the head of a softly wending path, built from '60,000 individual axe-hewn Portuguese granite setts'. We are invited to imagine that we are John Bunyan's Pilgrim, treading devoutly through 'Life', 'Death' and 'Spirit'. It is an undeniably moving experience – a holy way. There is no one else on the path and the English trees are dripping softly onto the American soil (we have gifted to the American Republic the acre of land in which Kennedy's spirit rests). Tennyson is not far away on this dank afternoon:

> The woods decay, the woods decay and fall,
> The vapours weep their burthen to the ground,
> Man comes and tills the field and lies beneath,
> And after many a summer dies the swan.

FROM 'TITHONUS'

The grove in which the Kennedy stone stands is a confused jumble of Classical, Celtic and Christian iconography. As well as an American oak, planted so that its red leaves will fall in bloody ruin every November to coincide with JFK's murder, there is a hawthorn set to one side as 'a symbol of Catholicism'. This may be so. After all, you can't get much more 'Catholic' than the Glastonbury thorn, which blooms miraculously every midwinter in honour of Joseph of Arimathea's visit to Britain carrying the Holy Grail – really, it does, for the perfectly good reason that it's a *Crataegus monogyna* 'Biflora', in other words it flowers twice a year. But the hawthorn is also an undeniably pagan tree, and an ancient symbol of power and fertility. 'Hawthorn blossoms have a potently erotic perfume which has been likened to the aroma of the sexual secretions of females,' as Jacqueline Memory Paterson memorably puts it in her book *Tree Wisdom* (1996). We really have gone out of our way to make JFK feel at home.

There's no doubt that the path is more magical than the grove, with its self-consciously significant trees. The rain has stopped and weak sunshine is now glancing through the almost bare branches. With no one around to ruffle the peace, the only sounds are some tentative birdsong and the traffic that is never far away in our busy land. A plane leaving Heathrow passes low overhead. It is one week since the terrorist attacks in Paris (I hope I never need to qualify that statement) and I am standing in Britain, on American soil, contemplating a president who was pointlessly killed in the year I was born. The world seems very small, all of a sudden, and shabby.

To cheer myself up, I head for the Ankerwycke Yew. It's a truly ancient tree, at least two thousand years old, and it lives in a meadow just across the river from Runnymede. The Thames has shifted a few times in the past thousand years, so it's entirely possible that King John had his nose pressed to the Magna Carta under the shade of its dark-green branches. The reason that JFK is memorialized here is of course because of that weighty document, the moment when a monarch was reminded that he couldn't just do what he wanted and that he had to take the rights of others into account. Power-hungry barons, yes, but not just them. The Magna Carta enshrined the idea that people had rights. The area is littered with oak trees, planted to celebrate the event, or to draw warmth from its power. I see that the Americans have been here again, digging an oak into soil taken from Jamestown, Virginia, to commemorate the bicentenary of their sacred Constitution; and the Queen dropped by in 1997, planting a rather more prosaic oak to mark National Tree Week. Eight English oak trees are growing around the commemorative Greek temple; another two have been added relatively recently by the Prime Minister of India and, for some reason, the Duke of Gloucester. There is something about these oak trees that is integral to the document and everything it represents. They speak to us of ancient freedoms, long predating the grudging concessions of a medieval king… and if one of them is currently lying uprooted and dying, well, we can always plant another one. Constant vigilance is all.

It is not so well known that there was also a Charter of the Forest (*Carta de Foresta*), this one signed two years later, in 1217, by King John's heir Henry III. It affirms and protects the historic rights of people to have access to the woods and royal forests, so they can carry on collecting firewood, feeding their livestock and foraging for food. It seems to me that this document is just as significant as its more famous cousin: after all, what did the barons care whether people could carry on profiting from woods? Quite the opposite, I'd guess. This one was genuinely for the common people. In November 2017, exactly 800 years after the first one was signed, the Woodland Trust and over seventy other organizations (charities, as well as the leading lights of the forestry industry) gathered in Lincoln Castle to sign a brand-new 'Charter for Woods, Trees and People'. It was an inspiring moment, and a timely reminder that we need our woods and trees just as much as we ever did.

The Ankerwycke Yew has stood on this same spot for over two thousand years. It's vast, of course, and seen from the right angle is almost like a yew forest or copse, rather than one tree. People have hung ribbons and small woven circles of wheat from its hollow, disintegrating trunk, in recognition of its power and in the hope that it can help them with their dreams or grief. Its branches loop upwards then drift to the ground; in some places they have even rooted again and are starting to form another circle of growing tree a few yards from the main trunk. The ground, as it is under all yews, is bare and scattered with fallen brown needles. There's a muffled empty thump when you stamp your feet: what goes on underground among the roots of a yew tree this old? Not far from the Ankerwycke Yew there are more yew trees, its offspring, some of them already a great age. If humanity could bring itself not to interfere for another thousand years, although there is little hope of that, then this tree and its descendants would all still be living. The same may not be true, alas, of the many box trees in the area, which can possibly already feel the distant but steady tramp of the approaching caterpillars.

The yew is constantly shifting, its tectonic bark stretching and breaking. If you peer into the dark depths of its empty trunk you'll see more peace offerings (and no litter), but also its thick roots lifted clear from the ground and groping for their next purchase. It's black in there – its top edges rimmed fungal white. And if you lean close to its trunk and look upwards, it's almost as though you're gazing at an ancient mountain range, with smaller trees growing from the rugged brown terrain. It's impossible not to think about age and death when you're in the presence of a yew. This one has not just witnessed the Magna Carta, but also the arrival of the Romans, the gathering of Saxon kings (King Alfred is said to have passed this way) and it's entirely possible that in the 1530s, under the branches of this very tree, Henry VIII had it away with Anne Boleyn, taking her roughly (we can only assume) up against its braced and trembling trunk. This tree can take us back to a time before pollution and the incessant whining of cars and planes. Who wouldn't want to be there, in the Tudor years? Sure, there was the strong risk of death by despots, disease and bad dentistry, but imagine the sweetness of the air. The deep, still forests and the clear blue skies. And the improbable silence, broken only by the urgent bellowing of the rutting king.

If I were to stretch for one final reason why Britain has more ancient trees than other countries, I'd say it's because we like our history – especially (but not only) the history that shows us in a good light. Magna Carta. Nelson. The longbow and the wooden wall. Of course, we are also famous for our world-class hypocrisy: we have saved these trees even as, at the same time, we have demolished the woods and forests that once surrounded them. But the survival of these few ancient trees is thrilling. We have passed them through the generations with veneration and love and they are still here, our friendly giants, looming in the woods, cities, hedgerows, farms and parks. We should seek them out and give them the protection they need. No, more than that. We should be on our knees at their feet, begging forgiveness and enlightenment. They have a wisdom that we need.

6

And mony ane sings o' grass, o' grass,
And mony ane sings o' corn,
And mony ane sings o' Robin Hood
Kens little where he was born.
It wasna in the ha', the ha',
Nor in the painted bower,
But it was in the gude green-wood,
Amang the lily-flower.

ANON,
'THE BIRTH OF ROBIN HOOD'

Full of Fame Is the Greenwood

Once there was an outlaw, and he lived in York, or perhaps it was Carlisle, or Rockingham or Nottingham, and his name was Robert (or Robin) Hod, or Hoode, Wood or Hood. He was a simple yeoman, working as a servant in the court of King Edward, but he stole some stores, or struck a man, and he had to flee to the forest of Barnsdale and there he lived out his doomed, fugitive days, deep among the trees, robbing or murdering unwary travellers and hoarding his plunder. Except, sometimes, he is said to have spared those who couldn't afford to be robbed:

> For he was a good outlaw,
> And did poor men much good.

ANON, FROM
A GEST OF ROBYN HODE (C.1450)

Perhaps it would be more true to say that he lived in Sherwood Forest, and he didn't rob or kill just *anyone* who dared venture down the forest tracks, it was the rich he attacked (with his Merry Men) and he gave the money torn from the purses of these wicked and tyrannical lords and bishops to the deserving and tyrannized poor. Except, by now it has emerged that Robin isn't really an impoverished yeoman, he is the true Lord of Locksley Manor (he's Robin of Locksley!), the son of the Earl of Huntington, and he's leading the Saxons in their righteous uprising against the Sheriff of Nottingham, Bad King John and the vicious Norman overlords who are looting, taxing and murdering the people of these oppressed and burning lands.

Here is something that we do know for sure: there never was such a wild and magnificent wood as Sherwood Forest:

> Hundreds of broad-headed, short-stemmed, wide-branched oaks, which had witnessed perhaps the stately march of the Roman soldiery, flung their gnarled arms over a thick carpet of the most delicious green sward.

> SIR WALTER SCOTT, FROM *IVANHOE*

And in this forest there were deep pools and forest glades, where otters played and the deer ran plump and free. And here Robin and his Merry Men feasted like no one has ever feasted before or since and drank deep from brimming flagons:

> There was feasting and rejoicing that night in the secret glade where Robin and his Merry Men did honour to their lovely queen. There was no lack of good roasted venison, great flagons of wine were set on the board, bowls of brown ale, and many another delicacy.

> ROGER LANCELYN GREEN, FROM
> *THE ADVENTURES OF ROBIN HOOD*

Because by now Robin has married his Maid Marian, his 'lovely queen', who may in fact be the daughter of the Earl of Huntington, or the ex-wife of the earl's son (because Robin is just a simple yeoman who fights and marries well), or perhaps she's a young woman lost in the forest who doesn't marry Robin (and he graciously doesn't ask) but she leaves the forest and becomes a nun. We don't know. But we do know, of course we do, that there never were such Merry Men. The laughter rang along the forest paths and soared up to the treetops (where Robin swung with infinite grace) and the sward was greener and merrier than anything before or since:

> What life is there like to bold Robin Hood?
> It is so pleasant a thing:
> In merry Sherwood he spends his days
> As pleasantly as a king.

ANTHONY MUNDAY, *FROM*
'METROPOLIS CORONATA' (1615)

And Robin could fight. In fact, he could kill with ease and pleasure. He could thrash the giant Little John with a stave (not always, but he hated to lose); he could slice and fillet the best-trained Norman knight with his sword (he never needed a shield); and he would even join the jousting lists, descending with awful vengeance on the best of King John's champions, shattering their shields and trampling their shabby foreign honour in the bloody English dust. But, more than that, there never was anyone who had such mastery of the longbow as Robin of Locksley. He could outshoot the finest archers in the land, even the Sheriff of Nottingham's best man-at-arms, and such was his skill that he could split his opponent's arrow in two, even when it had already been fired slap bang into the centre of the distant bullseye. The sheriff only called this tournament so he could flush Robin out of hiding, but to no avail. Robin and the Merry Men of Old England (Little John, Friar Tuck, Will Scarlett, Allan-a-Dale, Much the Miller's Son) are more than a match for these simpering, degenerate, authoritarian, Norman eunuchs and their cowardly, crossbow-wielding minions.

Robin and his men wear cloth of Lincoln green. They are green men, living among the trees, invisible to the town dwellers. They live in harmony with the plants and the animals and the soil; they have been absorbed by the land; they melt into the trees. Their weapons – the bow and its goose-feathered arrows, the sturdy stave cut from the nearest copse – are part of nature, and so are they. And even if Bad King John and the Sheriff of Nottingham send a mighty army, they will never find Robin, because the brambles in the woods, and indeed the forest floor itself, will rise up to protect him:

> Deep in the heart of Sherwood Forest, as the sun was setting behind them, Robin and his men came to a great glade where stood the greatest of all the forest oaks upon a stretch of open greensward with steep banks fencing it on either hand in which were caves both deep and dry. At either end of the shallow valley, and beyond the banks on each side, the forest hedged them in with its mighty trees, with oak and ash, with beech and elm and chestnut, and also with thick clumps of impassable thorns, with desolate marshes where an unwary step might catch a man or a horse and drag him down into the dark quagmire, and with brambles rising like high dykes and knolls through which even a man in armour could scarcely force his way.

> ROGER LANCELYN GREEN, FROM
> *THE ADVENTURES OF ROBIN HOOD*

Some people can find Robin. Marian does. And there's King Richard the Lionheart, who may be Norman, but he has befriended the Saxons and learned to govern and rule in harmony with the land and his peoples. And there's the sly, low-born bounty hunter Guy from the Yorkshire village of Gisborne, who dresses in the skin of a horse. Except, Guy is also high born – he's Guy de Gisborne, his first name pronounced 'ghee', like a tub of clarified Indian butter – and as the years pass the forest becomes a tangled noose of class distinctions. The Lady Marian's wimple flutters in the woodland breeze and she shrieks, helplessly, for Robin to rescue her from 'Ghee'. Or is Marian perhaps torn between choosing the outlaw Robin Hood and the undeniably handsome, and somewhat devilish nobleman, Sir Guy of Gisborne?

In the early days, when 'Ghee' is still 'Guy', the rough and fetid bounty hunter, Robin and Little John have a simple way of dealing with this fool, who is vividly described in an early ballad:

> A sword and a dagger he wore by his side,
> Of many a man the bane;
> And he was clad in his capull-hyde [horse-skin]
> Topp and tayll and mayne...
> 'I dwell by dale and downe,' quoth hee,
> 'And Robin to take I'me sworne;
> And when I am called by my right name
> I am Guy of good Gisborne.'

FROM *ROBIN HOOD*
AND GUY OF GISBORNE (MEDIEVAL)

For all Guy's swagger, Robin beats him in a sword fight, chops off and then mutilates Guy's head so that no one will recognize it, and disguises himself in Guy's rank and soiled horse skins. He then brings this gruesome bounty to the Sheriff of Nottingham and announces that he is Guy, returning with the head of the known outlaw Robin Hood. The sheriff is happily rejoicing when Robin rips off his disguise and rescues Little John (who had been captured earlier). The sheriff attempts to flee 'towards his house, in Nottingham town', but Little John takes careful aim and shoots him down with an arrow.

It doesn't end there, of course. It never does. There are many more ways to kill Guy/Ghee. And there are many ways for Robin to die. He is wounded and goes to find sanctuary in an abbey. But the abbess is having an affair with a Norman knight, and she poisons him, and then bleeds him to death, oh-so-slowly, drop by desperate drop. Or the knight creeps into Robin's room and stabs him when he is drained like a veal calf and weak from loss of blood. But Robin has time to kill the knight and fire off one last arrow – we all know that – and where it lands is where he is buried. Really, he should never have left the forest, his sanctuary and the source of all his strength. Perhaps he didn't. Perhaps he reclaimed Locksley Manor. Or he married Marian and lived in the greenwood, rejoicing and feasting, and rolling around the sward, for the rest of his merry days. He was, as Sir Walter Scott wrote, 'The king of outlaws and the prince of good fellows!' and he and Sherwood can never die:

> Robin Hood is here again: all his merry thieves
> Hear a ghostly bugle-note shivering through the leaves…
> The dead are coming back again, the years are rolled away
> In Sherwood, in Sherwood, about the break of day.

ALFRED NOYES,
FROM 'A SONG OF SHERWOOD' (*C*.1920)

In fact, and to the best of our knowledge, there really was such a man as Robin Hood. The details may be elusive – the forest, the Merry Men, Marian and the wretched Guy of Gisborne, quite where and to whom he distributed his loot, or what he was actually called – but it seems more than likely that some such outlaw existed. And there's no doubting the resonance of the legend. We like the idea that someone is out there, in the woods, taking money from the undeserving, high-taxing (but never tax-paying) rich and giving it all back to the grateful poor. It certainly bucks the trends of this or any other time.

'Gisborne' was the maiden name of my grandmother, Gwendolen, who married my grandfather, Gerard Fiennes, had five sons and so became known later in life (to some of us, at least) as 'Granny Fiennes'. She was a tiny woman, who dressed in Victorian black and ended her days in a small room in Tunbridge Wells, surrounded by great blocks of old-fashioned and impractical furniture transported from grander homes. She died before I was even five years old, but I remember our weekend visits, fraught with import, and a watchful old woman with claw-like hands, somewhat overwhelmed by the size of her own chair. My parents spoke slowly and carefully – and whispered outside Granny's door before we went in, which may have been the first time I realized that grown-ups do not always present an impenetrable, unified front. Perhaps Granny and I chatted at those moments, but it seems unlikely. What I do know is that for years afterwards I would happily play the part of Guy of Gisborne as I fought my friends with staves of wood; and I always sided with the dastardly Basil Rathbone in *The Adventures of Robin Hood*, even though he sneered and sulked and did not wear his tights with anything like the bulging manly panache of Errol Flynn. Even the famous illustrations by Louis Rhead, which make Guy of Gisborne look like a demented ass, could not dent my family loyalty. My father, you see, had told us – without a shred of evidence – that we were descended from 'Sir Guy' and so I knew he had to be good at heart.

No outlaw could hide in Sherwood Forest today – not for longer than a few hours – without being dragged out by the forces of law and order. About eight hundred years ago, Sherwood was a royal hunting forest stretching from Worksop in the north down to the city of Nottingham in the south, and it covered about 46,900 hectares: imagine a quarter of present-day Nottinghamshire covered in wood pasture, water meadows, heathland, wetland, hedgerows, fields and, yes, dense forest. For most of us, it's hard to visualize 'acres' and 'hectares', but if this helps, 46,900 hectares is the same as about 80,000 Premier League football pitches. Today, the National Nature Reserve of Sherwood Forest has dwindled to about 432 hectares; in other words, it's about one-hundredth the size it was in the days of Robin. Most of the early destruction was done by the agricultural and industrial revolutions (this is – or was – coal country), but the old forest was predictably pulverized by its familiar twentieth-century enemies: the Forestry Commission, two world wars and a booming population.

Still, some of it remains, and the Sherwood Forest Trust is working to restore what it can. Thousands of native trees have been planted, new ponds and wetlands are being created and copses, hedgerows and grasslands are being renewed. They are even planting 'heritage variety orchard trees' in the local farms. I've arranged to walk here with Sian Atkinson, senior adviser at the Woodland Trust's conservation team. Not that the Woodland Trust is responsible for Sherwood Forest, but it's certainly an iconic and resonant place in which to meet, and the plan, she tells me, as we stroll around its lovely, battered remnants, is to work at the 'landscape level'. This is the guiding principle of all present-day conservationists. The idea is to create 'corridors' between the remaining native woods – hedgerows, certainly, but also rivers, scrub, moors and heath – along which wildlife and plants can travel and spread. Sometimes these 'corridors' do not even have to be connected, but might in reality be isolated 'stepping stones' of copses, ponds or large trees, rather than one continuously linked sanctuary. The important thing is that they are there, enabling the flow and exchange of genes.

If, once again, you can bring yourself to imagine Britain as a garden (but this time from the perspective of the native woods and everything that lives in them), then what we have now is a scorched desert of chemical-soaked farmland, grasslands patrolled by voracious sheep, inhospitable towns, impassable roads, toxic landfill, the odd disused mine, some factories and car parks, conifer plantations and, dotted here and there (mainly in the south), a few broadleaf woods. Somehow, we need to make it possible for wild nature (the pollen, seeds, nuts, orchids, spores, beetles, ants, birds, toadstools, badgers, voles, weasels – the whole infinite variety of living organisms) to travel between the woods, sidestepping the sheep and the pesticides, without starving or falling under the silent wheels of a Toyota Prius. The inspiring aim is to make sure that species don't become isolated and that every ecosystem remains or becomes as richly diverse as possible. It's time to back off and learn to allow unadulterated nature to encroach on our lives. In fact, while we're at it, now is probably also a good moment to stop pretending that we're in charge. We're part of this thing, and with climate change and wild weather becoming a racing certainty, we should unlock the cages and let everything spread out and take cover. There's a storm brewing and we're going to need all the native diversity we can muster.

Sherwood has always been famous for its oaks. In 1790 a man called Hayman Rooke made a survey of the timber in the forest or, as he called it, 'A Sketch of the Ancient and Present State of Sherwood Forest, in the county of Nottingham', and ended up wringing his hands over the wholesale destruction:

> It appears that this Forest, so late as the beginning of this century [i.e. 1700], was full of Trees, the Rev. Doctor Wylde, Prebend of Southwell, and Rector of St Nicholas in Nottingham, assured me he had often heard his father, William Wylde, Esq. of Nettleworth, who died in the year 1780, in the 83rd year of his age, say, that he well remembered one continued wood from Mansfield to

Nottingham. Since that time the Forest has been pretty
much cleared.

Feelings of loss multiply down through the generations, but our lives
are too short (even the venerable William Wylde's) to grasp what is
happening. There are still about 1,000 'veteran' oak trees standing in
Sherwood (down from 49,909 good-sized trees in 1609), specimens
that are at least five hundred years old, although about half of these
monsters are now hollowed out and dead – of old age and thirst,
brought on by a shrinking water table. Apart from the 'veterans',
there is an unfortunate generation gap, marking many centuries
when all the useful timber was felled for the furnaces and the navy,
and before there was a new wave of planting that took place 150
to 175 years ago. When surveying the forest, or planning for the
future, we have to think in a 'tree time' of centuries. The 'Major
Oak' is the oldest of them all and, according to a rather optimistic
note from the Sherwood Forest Trust, it is the most famous tree
in the world. Maybe it is: although the General Sherman, a giant
redwood in California, and the world's largest tree by biomass,
would surely run it close. And what about the cedars of Lebanon?
Anyway, the Sherwood Forest Trust is now planting frantically,
with one eye on the year 2025 (when it hopes to have increased
the forest's 'characteristic habitats' by sixty per cent) and one eye,
we trust, on the year 2500.

The Major Oak will surely be gone by then. The best estimate
is that it is now over a thousand years old, although some put it
at 1,200. Whatever the truth, it was undoubtedly born before the
Norman Conquest, and even if that means that it was too young
for Robin and his Merry Men to give it more than a casual glance,
it was still *there* when they were flitting through the glades. It used to
be known as the Cockpen Tree (because of the cock fighting that was

staged in its shade), but since 1790 it has been renamed the Major Oak after the very same (Major) Hayman Rooke who laboriously itemized the dwindling oaks of Sherwood. It is now fenced off (to stop visitors from pounding down the soil) and its immense branches are supported by a series of stout wooden crutches, giving it the air of a sunken, Falstaffian old man who might, at any moment, lurch from the earth. Some people grumble that this late-life nursing is taking up too much time and money – why not just let nature take its toll? – but who would want to be the one to pull the plug on this granddaddy of the woods?

Sherwood today is not the Sherwood of our dreams. The paths are well trodden by horses, dogs and people; the undergrowth is cleared (although it is being allowed to return); and much of it feels like well-worked, lightly wooded parkland. There is no chance that Robin Hood would choose this ravaged forest as a refuge, although he could probably mingle unobtrusively with the crowds in 'The Forest Table' restaurant. As E.M. Forster grumpily wrote in 1970 in an endnote to his novel *Maurice*:

> There is no forest or fell to escape to today, no cave in which to curl up, no deserted valley for those who wish to be left alone. People do still escape, one can see them any night at it in the films. But they are gangsters and not outlaws, they can dodge civilization because they are part of it.

He's right. A fugitive from the law wouldn't choose a modern British wood in which to hide, especially if she or he had a bunch of Merry Men to feed and clothe. The runaway in Geoffrey Household's *Rogue Male* (1939) did make for the woods, burrowing deep into an old rabbit hole that he'd enlarged with his teeth (or something – it's a very *odd* book), until the Nazis run him to ground; and Richard

Hannay in *The Thirty-Nine Steps* (1915) dashes between the Scottish woods and moors, also pursued by evil Germans and their Fifth Column sympathizers. But these tales are fiction. Wild Britain has been lost, and whatever you may hear from Bear Grylls, there's really no point in learning how to build a trap for voles or an oven for roasting earthworms, except for your own amusement. You are never more than a brisk walk from a hot cappuccino and a bacon sarnie.

In the 1940s John Stewart Collis worked in Cranborne Chase in Dorset, where Thomas Hardy once roamed and wrote, and describes in *Down to Earth* how he thinned the small wood single-handedly, using only an axe and other hand tools. As he struggled with the rampant ash saplings, he decided that if he were ever on the run from the law, this would be the perfect place to hide out – until he learned that in the previous century 'thieves, murderers and criminals of every grade' had had the same idea – and so the entire area was disafforested. His little wood was the last, sad remnant of a great forest refuge.

It has been a wearily common theme throughout history. The authorities don't like forests, because they don't like places where people can hide. The Americans napalmed the jungles of Vietnam and showered them from on high with the defoliant Agent Orange. The Europeans razed the great forests of North America (but only after their predecessors had done what they could with more primitive tools). Edward I would have done the same in Wales, if he'd had the technology. Instead, in 1277 he ordered all woodland to be cleared that was less than the length of a bow shot from the main roads; and in the year 1287 more than three thousand English woodcutters, charcoal burners and diggers were ordered into Wales to slash and burn anywhere that might shelter an enemy or an outlaw. The same destruction was visited on the pine forests of the

Scottish Highlands after the 1745 rebellion. It meant the rebels had nowhere to hide – and there was the handy side-benefit of a tasty load of timber and a big fat profit for the absentee landowners. Two hundred years later their descendants would be gunning down grouse in a denuded landscape. As Forster says, if you want to hide today, you need to head for the city. The rest of the land has been scoured clean.

Here's John Keats's 'Robin Hood':

> And if Robin should be cast
> Sudden from his turfed grave,
> And if Marian should have
> Once again her forest days,
> She would weep, and he would craze:
> He would swear, for all his oaks,
> Fall'n beneath the dockyard strokes,
> Have rotted on the briny seas;
> She would weep that her wild bees
> Sang not to her – strange! that honey
> Can't be got without hard money!

The only fugitive of note in modern times who headed for the woods (so far as we know) was the miserable Raoul Moat, who shot himself in a ditch outside the village of Rothbury in Northumberland in July 2010, surrounded (at a safe distance) by armed police, his best friend Tony Laidler and – sporting a crumpled dressing gown – the ex-England footballer, Gazza, who was trying to coax him out with an offer of chicken and lager. Moat had been on the run for a week, having shot and wounded his ex-girlfriend, killed her new boyfriend and blinded an on-duty, unarmed policeman. As Moat ran, and hid, excitement bubbled across every news platform and branch of social media. Ray Mears, the survivalist and woodsman, was brought in to track him, but people seemed to be hoping that Moat would carry on evading capture for as long as possible. Sympathy grew for the hunted man. There was, obviously, nothing remotely heroic about Moat, but within hours of his death a Facebook page had been created called 'RIP Raoul Moat You Legend', with thirty thousand followers, many leaving heartbroken messages of love and respect. These events (and the almost instant elevation of Moat to the status of 'legend') now shed a rather uncomfortable, retrospective light on the more glorious adventures of Robin Hood – and it's for this reason that I'm going to stick with the belief that it was Sir *Ghee* who was the real hero all along.

The last time a longbow was used as a weapon of war was early in the Second World War, when Lieutenant Colonel 'Mad' Jack Churchill ambushed and shot dead a German sergeant with the yew bow that he'd brought from home. Imagine that – and think how the German must have felt in his last moments, having quite possibly survived tanks, artillery, snipers, machine guns, aerial bombardment and grenades, only to fall victim to some medieval high jinks. He must have been dumbfounded. Of course, the longbow was never a joke. In its day it was a weapon of terror, a technological advantage as potent as the rifle or the cannon, used by the English to win the battles of Crécy, Poitiers and, most famously, Agincourt, when the arrows were so many and so thick that they darkened the sky. The arrows are about thirty inches in length and the fletchers stuck on the guiding feathers with a viscous, sticky sap taken from bluebells. It would be true to say that the longbow is a woodland creation. As we all know, the best bows come from the yew tree, although they can also be made from the wych elm if you've already chopped down most of your yews, which is – inevitably – exactly what happened. Despite what I once imagined, the yew is not by nature a solitary tree, confined to churchyards and parks. There were once yew woods and forests all over northern Europe, but they were destroyed by the demands of the arms industry. In just one year (1341), when the Hundred Years' War was raging, the government alone ordered 7,800 bows and 13,000 arrows – a huge quantity of wood. Once the yew's value (or scarcity) became recognized, kings started to take more care to protect it; but it has always been considered bad luck to chop down a yew tree, even though since Neolithic times it has been used to make bows. There is something that shifts and broods in its shade that you wouldn't want to disturb – and its real message, or so said Sylvia Plath in 'The Moon and The Yew Tree', is 'blackness and silence'.

Despite the ravages of war, it is still possible to visit a yew wood in Britain. On a hot blue day in early October I am walking among yew trees, nothing but yew trees, in Kingley Vale on the slopes of the chalk downs of West Sussex. The yew is one of those trees that suffocates other species, rather than tolerating (or even welcoming) their presence. It's rather like the beech in this way – give the yew enough time, in the right place, and it will inexorably grind out its own dominion. You can see it happening here. There's a sizeable oak, caught and surrounded by yews, and it has been starved to death – of light, perhaps, but also (and mainly?) of water. I imagine the yew is a greedily thirsty tree. Certainly, the ground here rings hollow and dry, and it's coloured a dusty light brown, rather like the trunks of the yews themselves, although when a snatch of pale sunlight breaks through the canopy you can see a faint pulse of pink in the bark. At the front line, where the yews meet a mixed woodland, there are anemones growing under the placid shade of an oak tree – but behind me, under the yews, there is nothing but dust. It is said that the yew will connect you to the other world, but unlike the alder, which can steer you back and forth, and whose leaves were once used to soothe weary medieval feet, the yew will only show the spirit-traveller a vision of the void: 'blackness and silence'. It is a powerful tree, and powerfully poisonous to humans, especially the stone of the fruit. Perhaps that's what the ancients meant when they spoke of a one-way journey in the presence of the yew. In classical times the yew was sacred to Hecate, the goddess of magic, crossroads, ghosts and the moon, although she could bring good luck to any family that treated her well. In *Macbeth* her cauldron contained 'slips of yew, sliver'd in the moon's eclipse' – and it was bad luck to carry a piece of yew that had been blown ('sliver'd') from the tree. Basically, you couldn't pick it up and you couldn't chop it down.

It was certainly considered a very bad idea to fall asleep under a yew tree. They move, too. The trunk will reach a certain height, about fifteen metres, and then will grow only outwards – hollowing as it swells. The branches dip and brush the ground. In places, they will root again, and rise again, and the yew tree will live for an age. Well, not quite 'an age': the lifespan of three yews is the same as 'an age' – and we are told there are seven ages 'from creation to doom'. The stave of the longbow was taken from the trunk, cut so that about one-third of it was 'sapwood' and the rest 'heartwood'. The finished bow could carry more tension than a bow taken from any other tree and I vividly remember the moment in Sir Arthur Conan Doyle's *The White Company* when the English archer shoots his arrow clean through a shield that has been left as a target, while the Frenchman's crossbow bolt simply sticks and hangs, limply, in the outer layer.

There must, surely, have been many yew trees growing in Sherwood Forest when Robin Hood hid in its depths. He needed the tree for his bows (his staves would have come from the quick, clean-growing ash), but the yew is also an incomparably good hiding place. People don't often look up when walking through woods – they have their eyes on the ground, scanning for mushrooms, flowers and rabbit holes, or on the trunks of the trees – but if you sit under three of the huge yews here in Kingley Vale, the ones known as the three witches, you'll see what I mean. Their branches twist and flail, shade and sunlight mingle confusingly, and a small number of green men could easily hide in the upper branches. The yew is also evergreen, which helps the winter outlaw. One of the 'witches' has imploded, but the branches of the other two are thickly inter-twined. It is sisterly – or reminiscent of the doomed lovers Tristan and Isolde, whose separate graves sprouted yew trees that leaned, and reached, and clung to one another, despite the efforts of the cuckolded husband to keep them apart. If you look up here, 'E' has engraved that she loves ('hearts') 'A', but the 'A' has since been crossed out. Does Eve no longer love Adam?

I remember playing 'hide and seek' at the strange little school I was sent to aged eight, and having to go and find our scout master in the woods. Apparently I walked underneath him at least three times, he told me, crowingly, when the game was over and no one had found him. He'd been hiding in the upper branches of a large yew tree – and, he said, you should always remember to look up when looking for people in woods. (I just worried what I'd been doing when he was gazing at me from the gloom of the yew. Was I talking to myself?) Mind you, he was a peculiar man. He taught us geography with a stubbed-out cigarette butt, secured by strands of yellow phlegm, dangling from his lips throughout the lessons, and it just needed one well-aimed question about 'the war' to set him off on his favourite story, about how he'd commanded a 'little tank' that had once met a 'big tank', and how he'd driven round and round this 'big tank', firing all the time, while the 'big tank' rotated its gun in an impotent and futile attempt to swat him away, until finally the 'big tank' had given up, or exploded, or peace had been declared. Come to think of it, the ending of this interminable story always was unclear (perhaps he never reached it?) but he would fill the blackboard with swirling chalk diagrams of Celtic complexity (the little tank whizzing round the big tank), and wheeze with intense faraway laughter until the lesson ended. I learned nothing about geography, of course, but I did learn about hiding in yew trees.

> There is a Yew-tree, pride of Lorton Vale,
> Which to this day stands single, in the midst
> Of its own darkness, as it stood of yore;
>
> WORDSWORTH, *FROM* 'YEW-TREES'

The yew is a warlike tree, bringing visions of darkness. During the Second World War the Spitfires used to practise firing all along this valley and it's said that several of the yews were blown up, or still bear the scars of the bullets. There's no sign of that, but the great trunks are lipped and furled and could quite easily have absorbed any amount of armaments. I'm struck by how different a yew wood is – in atmosphere and appearance – from a larch or conifer plantation. Most obviously, the trees are not planted with any regularity, and they are not of a uniform age (giants mingle with striplings), but there's more to it than that. For one thing, sunlight reaches the forest floor (even though nothing dares to grow on it), and that lifts my mood. But there's a peace here, a sense of spirit and wonder that is never present in a plantation. I could sit under these trees for hours, whatever the ancient writers may say, and if I slept I'm sure my dreams would be strange and Dalí-esque, the contorted, elephantine trees nuzzling and probing at my subconscious, but

I would not be afraid. At least, that's what I think on this pleasant, sunny October morning. I'm not entirely sure I want to put it to the test at the next new moon.

When you leave the ancient yews behind, and emerge at the foot of the chalky downs, you can see that the descendants of these ancient trees are now spreading up the hill and eating away at the grassland. I love woodland more than most, but it's not the only habitat that is threatened by twenty-first-century life. Grassland is another – and its survival depends on man and his grazing livestock. It's the temporary cessation in sheep grazing that has allowed the yews to spread so vigorously up the hill. This is something that would have intrigued Sir Arthur Tansley, whose memorial stone stands at the top of the sun-bleached hill:

IN THE MIDST OF

THIS NATURE RESERVE

WHICH HE BROUGHT INTO BEING

THIS STONE CALLS TO MEMORY

SIR ARTHUR GEORGE

TANSLEY, F.R.S.

WHO, DURING A LONG LIFETIME,

STROVE WITH SUCCESS

TO WIDEN THE KNOWLEDGE,

TO DEEPEN THE LOVE AND

TO SAFEGUARD THE HERITAGE

OF NATURE

IN THE BRITISH ISLES

NOVEMBER 1957

Well, who wouldn't want that as their memorial? To strive with success… to deepen the love… I can only assume that Sir Arthur is resting easy in his grave at the end of a long life well spent. Among his many other achievements, Sir Arthur gave his name to the 'Tansley Model', which is one of the two key theories of woodland development, and is based on his assumption that at one time the whole of Britain (perhaps the whole world) was covered in thick woodland, up until the moment when huge herbivores and then humans arrived and started browsing and burning, digging and cultivating, building and bulldozing. And so, over many millennia, we loosened the trees' grip on the landscape, making more and more clearings in the forest until Britain had become one big clearing, surrounding a few pockets of surviving woodland. If you've ever played the utterly addictive computer game 'Age of Empires' you will know that the way the little Greek woodcutters set about clearing the screen of forests is based on the Tansley Model. It tallies with our instinctive belief that we are living in a world that once belonged to the trees – and it is demonstrably true, surely, that there were once deep forests where we now live, work and shop. But it may not have been that simple. The alternative view was set out by Frans Vera in his book *Grazing Ecology and Forest History* (2000) – in which he suggested that ever since the last Ice Age the northern hemisphere has been a mix of open space, wood pasture and forest. And that even if humanity were to disappear, taking all of our wretched sheep with us, there would still be a mix of pasture and woodland, kept clear by herbivores and shifting, inhospitable soil. The natural state of Britain is not thick woodland from Land's End to John O'Groats, but a mix of moorland, grassland, heath, field and woods. How else do we explain the preponderance of non-woodland specialist plants? Needless to say, the academics have not yet agreed about any of this, although I think it's possible that the pendulum is now swinging back towards Sir Arthur and his Sleeping Beauty Forests. I do hope so.

Before reaching the top of the hill above Kingley Vale, there is one last path through the yew trees. I realize I have been wrong about the differences between conifer plantations and yew woods. They both carry darkness. The spread of yews on this hillside is relatively new, the trees are all the same rough age (they're probably all less than a hundred years old), and they are even growing in shaky rows, almost as though they have been planted. Perhaps they were. The floor of the forest path is reddish brown and smeared with pale sunlight, and so long as I stay on the path I can see lattice strips of muffled blue sky above the reach of the enclosing trees, but it is very dark in the woods on either side. I have a cold conviction that I should stay on the path. Ahead of me, a bare, dead yew has fallen from the woods and is crouched like an immense stag beetle on the path's bank. To either side, there is no life other than the silent yews; no understory of bracken or brambles; no birds singing. At the edges of this path, where the dim light struggles to be felt, the trees are weighed down with clusters of ripe red berries. They look soft, plump and heavy, ready to drop into someone's open mouth. They look like they would burn a hole right through it.

It is a real relief to leave the woods and reach the bare, open top of the grassy downs. It is still a surprisingly bright blue day. I turn my back on the yews and gaze out over the fields and small woods to the north. If I were Sir Arthur, I think I'd pull out a pipe at this moment, knock out the old ash on the triangulation point, tamp down some new tobacco, fire it up and confirm my own theory: of course this was once all woodland. Just look at the way the woods are kept at bay by the busy workings of humanity – the fields and fences, roads, buildings and farms, the crops and hungry livestock. Given their chance, the woods will return.

And yet, it's impossible to spend any time among the trees and not think about their fragility in the face of all this human busy-ness.

Only a couple of centuries ago it might have been possible to imagine that the woods – the whole natural world – had a potency and a strength that was immune to the relentless march of humanity, but surely no one can think that now. We are too powerful. Many of our leading nature writers – Naomi Klein, George Monbiot, Richard Mabey – have all said that the world is strong and will find its way back, we just need to give it a bit of breathing space and some much-needed recovery time. I'm sure that's true. But do you honestly see it happening? And if so, when? The real truth is that we're high on petrol fumes and sugar, we've got our hands on the world's throat, and we're not going to let go until we've shaken it down and choked the life out of the wild and unruly bastard. How can this possibly end well? There is no doubt that something will survive – and some people seem to find it reassuring that nature's strength can be measured in the survival of lichen and cockroaches. And maybe, in a few million years, once the chemical dust has settled, another bunch of hungry apes will emerge from another regenerated African forest, and the whole damn thing will start all over again.

The yew really does spread darkness. But, still, it is impossible to spend time in the woods – any woods – without wondering about much more than just the glory of the trees. Of course, they are heart-achingly beautiful. In Japan, people are taken *Shinrun-yoku* – 'forest bathing' – to walk among the woods and cure them of any number of mental and physical ills. The recovery time of patients in hospitals can be greatly reduced if only they are given a view of the trees. The simple act of plunging your hands into the soil of a forest can alleviate depression: it's thought that traces of hope (there's a scientific word for it…) enter the bloodstream through tiny cuts in the skin. We walk in the woods and think of other things. How best to preserve them and create more. How to save them from the threat of disease, pests, alien species, rampant rhododendrons and muntjac; mankind's mad and wanton destruction of his own home – his seeming hatred for isolated beauty and loathing of places of calm and seclusion, his lust to own and acquire more. The growth of everything: towns, cities, roads and runways, the population, the carrier bags and all the pointless tat and crap that no one needs and never wanted. If you spend time in the woods, it's impossible to avoid the biggest questions of all – what's it all for? Not just the woods, but everything. What do we think we are doing? What on earth is the point? Wouldn't it be better if none of us were here?

Sorry. I do hope you're not expecting any answers.

7

Oh, I love
To sit me there till fancy weaves
Rich joys beneath a world of leaves

JOHN CLARE,
FROM 'WALKS IN THE WOODS'

A Vision in a Dream

Let's leave the yews. Woods are places of magic and delight, as Rudyard Kipling knew well:

> Of all the trees that grow so fair,
> Old England to adorn,
> Greater are none beneath the Sun
> Than Oak, and Ash, and Thorn...
>
> Oh, do not tell the Priest our plight,
> Or he would call it a sin;
> But – we have been out in the woods all night,
> A-conjuring Summer in!
> And we bring you news by word of mouth –
> Good news for cattle and corn –
> Now is the Sun come up from the South,
> With Oak, and Ash, and Thorn!

KIPLING, 'A TREE SONG'

I'm visiting Bateman's, near Burwash, Kipling's home in the folds of the ancient Sussex hills from 1902 to 1936 and the dreamily resonant place where he first drew that poem on to the page. It appears early in his book *Puck of Pook's Hill*, a rather dated attempt to inspire a love of English history among younger readers, although if nothing else he should know and rejoice that he succeeded with Simon Schama, who has said it was his favourite book and that 'for a small boy with his head in the past, Kipling's fantasy was potent magic'. The book tells the story of Dan and Una, two children who live and play in these hills and who meet Puck, 'a small, brown, broad-shouldered, pointy-eared person with a snub nose, slanting blue eyes, and a grin that ran right across his freckled face'. Dan and Una were performing 'as much as they could remember' of *A Midsummer Night's Dream* after tea on Midsummer's Eve (as all children surely did in those long-lost days) in a small, natural amphitheatre, 'overgrown with willow, hazel, and guelder-rose', when the real Puck appears and joins in. The magic, which I took to heart, was sparked by the children performing the play *three times* on Midsummer's Eve, in the middle of *a Ring*, right under one of the *oldest hills* in Old England. Over the next few chapters, Puck transports Dan and Una back into the past, to witness the key moments of early English history: the Normans, the Saxons and the Vikings, the Romans at Hadrian's Wall, and up to the Middle Ages and the happy days after the signing of the great 'Magna Charta'. But what lingers longer than any of the history is the smell and the taste of the Sussex hills and its wooded valleys, brought nonchalantly to life by Kipling – and, above all, the magic that flows through every hawthorn and ash tree and every flower in the languid summer meadows.

I remember the first time I came here, walking along a crooked path underneath a strip of hazel coppicing just beyond the gates of Kipling's garden. It was June, near Midsummer, towards the end of the long day, and no doubt Puck was fidgeting in some recess of my mind, but I remember the way the light changed in the treetops, and that I was sure there was something that might be half-glimpsed through the shifting leaves, something hidden in a space between the sunlight and the shade. I've had this feeling on paths in woods many times since, in every season, when a fleeting eddy of wind on the way ahead lifts and drops a few fallen autumn leaves, or a faint shaft of light briefly catches the drifting tufts of radiant meadowsweet.

Something disappears. It's always just out of reach, or out of focus, at the far end of the path; or it's behind you, when you glance back, and there's a ripple in the canopy in the very place you were standing just a minute before. If you gaze long enough at David Hockney's paintings of Woldgate Woods, a series of works that stretch from May to December, you can see it there, on the path and in the flicker of the leaves. And it was here, on this path in June, which had now become a darker place, waiting in silence on the edge of something unfamiliar. The undershrub at the fringes of the path, the brambles, nettles and foxgloves, had risen chest high from the thick earth; the grasses, pregnant with seed, were around my knees; and when I looked up it was as though the top branches and their leaves were drawing themselves at that very moment across the last few strips of the dying blue sky. I was swallowed up by the wood.

It was the height of summer then, long ago, but I'm here again, in late November, on the same path, just beyond the mill and the pond, and I'm looking at the same dabs of gold on the same muddy brown earth, only this time the leaves are dead and falling. There's no denying that there is something I want to call 'magic' in the air. Of course, Schama would say – does say! – that 'Landscapes are culture before they are nature; constructs of the imagination projected onto wood and water and rock.' And no doubt I'm just projecting Puck back into his familiar Sussex trees. But at the very least we should consider this:

> It is an idiosyncrasy of our culture that belief in an invisible supreme supernatural being is a respectable part of religious experience, while communicating with trees may be taken as a sign of mental disorder.

LUCY GOODISON, FROM
HOLY TREES AND OTHER ECOLOGICAL SURPRISES

Once upon a time people worshipped in the woods and venerated the trees. There were faeries and elves, imps, pixies and little people, centaurs, fauns, sylvani, nymphs, Cyclops, dryads and satyrs; there were wood creatures of every kind, playing in the glades and living in the trees. Often, they *were* the trees. Merlin gathered his mistletoe from the heavy-branched oaks and Pan haunted the wild forests and the remote mountain paths. There were Druids who ruled the people and protected their sacred trees; and the gods they revered feasted on bloody sacrifices in the inviolate groves. Long ago 'the god inhabited the tree or sacred stone not in the sense in which a man inhabits a house, but in the sense in which *his soul inhabits his body*' (my italics, Robertson Smith, *The Religion of the Semites*, 1889).

Imagine that there were still nymphs and dryads living in the trees, living *as* trees, just like they did three thousand years ago in the time of Homer. And…

At their birth pines or high-topped oaks spring up with them upon the fruitful earth, beautiful, flourishing trees, towering high upon the lofty mountains (and men call them holy places of the immortals, and never mortal lops them with the axe); but when the fate of death is near at hand, first those lovely trees wither where they stand, and the bark shrivels away about them, and the twigs fall down, and at last the life of the Nymph and of the tree leave the light of the sun together.

HOMER, *FROM* 'HYMN TO APHRODITE'
(TRANS. H.G. EVELYN-WHITE)

These lovely Homeric nymphs lived wild in the mountain forests, their fate umbilically linked to the pines and high-topped oaks that had sprung out of the earth at their birth. When they died, their tree also withered, but the reverse was also the case – and one very good reason, if any were needed, why mortals really should not cut them down with an axe. Lord Byron nursed a similar belief. He'd planted an oak at Newstead Abbey, his ancestral home, when he'd first visited the place (he didn't linger long in its damp and gloomy halls) and no doubt he would anxiously and regularly check on its health from the fleshpots of Istanbul and Mayfair. We know that he died of a violent fever, on his way to save the Greeks from Ottoman oppression, but I have no idea what happened to the tree. Perhaps it's still there.

Right now, I would find it a relief to give in to the siren calls of these ancient beliefs. The weight of our history and culture is against us, but it's hard to be always dismissing them. In fact, I'm not even sure why I try, other than a reflexive cringe of English embarrassment. In the eighteenth century the English were known for their absurd emotional incontinence, weeping at every opportunity, and their almost hysterical attachment to daemons, faeries and all sorts of supernatural cant, unlike, for example – and can this really be true? – the calm, stoical Italians. At this moment, in Kipling's woods, I could believe anything. I am following the path through the hazel coppice. It leads to a field, and a gate, and then another field that rolls uphill to a windblown copse. Is this where Dan and Una met Puck? There are woods as far as I can see, and there's no sign of any human activity, other than empty fields and frayed hedgerows, and a distant church, purveyor of a different kind of magic, its steeple poking through the treetops. Leaves are falling all around in thick gusts, even though I'm standing in the middle of a field, three hundred paces from the fringe of the nearest wood. I'm troubled by the superstition – my entrenched belief – that I have to catch twelve leaves before the end of December if I want to ensure good luck for the following year. There are still two to catch and the branches are emptying fast. It has become very important that I do this today: I left it too late the previous year and things were not always easy. It's ridiculous, of course, but the idea has taken root. Even so, it is not easy to catch falling leaves in a windblown field. They buck and twist erratically, especially these corkscrewing oak leaves, and I'm aware that I must look absurd, alone in the middle of this field, lunging and flapping at the air – and I still haven't caught anything.

There's a pond in a clearing in the copse at the top of the hill. The earth is shallow here, rocks showing through the damp leaves. It feels old, and the water in the pond is dark and very quiet, although the brambles under a wild cherry tree are pulsing with green life, even this late in the season. As soon as I arrive in this copse, one spent oak leaf circles down, slowly and easily, and I catch it. And then another one follows, and I grasp it hungrily, guaranteeing a happy, healthy year. And then, only a moment later, another leaf drops (or is placed) into my hand and sticks there and it takes me several frantic moments to shake it loose. Thirteen leaves… that cannot be good news. In a way, I'm glad that Puck has not lost any of his impish humour.

At the far edge of this copse there's another field, although this one is thickly spread with young, self-seeded oak saplings, spreading outwards with military intent. I think of Macbeth and the wood that moved, another bit of magic that seemed impossible until he looked out from the walls of Dunsinane and saw that his world was coming to an end. Once, we saw magic in every moving tree and silent flower. We all knew that the reason cowslips have red spots is because the fairies have removed the freckles from a sleeping girl; and that the rowan tree – the mountain ash – would keep the witches at bay. People would plant it in their gardens just to make sure:

Rowan tree and red thread
Haud the witches a' in dread.

There's one growing just here, on the edge of the small copse, its berries hanging coral red in the autumn light. I'm briefly tempted to tie a few twigs to the collar of my dog (it's a well known fact that this will increase the speed of your hound), but I reckon the squirrels could do with some respite.

The elder was a more ambivalent tree. The Greeks would scrape the soft pith from its branches to make flutes, and now we gather its berries and flowers for wine, jelly and cordials; but it was known that Judas had hanged himself from an elder tree, or so says Biron in Shakespeare's *Love's Labour's Lost*, and witches were said to frequent this ragged half-tree and draw strength from its berries. Then again, the elder could be used to undo evil magic, cure toothache and fevers, induce a peaceful sleep and, with a twig in your pocket, you would be less likely to commit adultery (should you feel the need to quell the urge). If you do find yourself threatened by a witch, then the extract of ferns is thought to produce invisibility, which also makes it useful for spying on furtive lovers. And when is kissing in season? When the gorse is in bloom.

I am drunk with the honey wine
Of the moon-unfolded eglantine,
Which fairies catch in hyacinth bowls.

PERCY BYSSHE SHELLEY, *FROM*
'WINE OF THE FAIRIES'

The hawthorn, as Kipling well knew, is the meeting place of faeries (with that dangerous middle 'e'), so you should never fall asleep under a hawthorn tree on Midsummer's Day, for fear of being spirited away; and nor should you, on any account, carry its flowers into your home. Ah, but then again:

> The fair maid who at the first of May,
> Goes to the fields at break of day,
> And walks in dew from the hawthorn tree,
> Will ever handsome be.

The most famous hawthorn in Britain is the Glastonbury Thorn, which 'blossoms at Christmas, mindful of our Lord'. The early Christians decided this was because it must have sprung from Joseph of Arimathea's staff when he rested it on the ground one Christmas Eve. The remnants of this tree were visited by Celia Fiennes in the late seventeenth century on her horseback journey around Britain, riding side-saddle. She found it in the ruins of Glastonbury Abbey, growing on a chimney: 'this the superstitious covet much and have got some of it for their gardens and soe [*sic*] have almost spoiled it'. It was just as well that 'the superstitious' had been clipping and rooting the branches: the Glastonbury Thorn has always attracted vandals and it has been attacked many times over the centuries, by Puritans during the English Civil War and, most recently – shockingly – in December 2010 by some pillock with a chainsaw. Despite these assaults, its successors live on. The early Christians were urged to make use of the places of worship they found and supplanted – the rocks, groves and sacred trees. It was easier to absorb the pagan ways than battle them and it's why many yew trees seem to predate the churches that now slumber next to them.

There wasn't a tree in the woods that couldn't be tapped for its magic. Mélusine Draco in *Traditional Witchcraft for the Woods and Forests* says of the willow: 'To create a charm to attract prophetic dreams, visions and inspiration place a sliver of bark in a pouch of pale green fabric, and place beneath the pillow during sleep.' Then again, the Israelites hung their harps on its listless branches as they wept by the rivers of Babylon. It was a tree of bitter sorrow (try licking the inside of the bark…) as well as prophecy. Also, you should never chop down an alder tree without permission from the tree. 'To the Celts,' writes Yvonne Aburrow in *The Enchanted Forest*, the alder 'was a faery tree of divination and resurrection, generally associated with death, the smith's fire, and the power of evaporation.' When it is cut, the alder bleeds red (which spreads, briefly, into its lovely white wood), and so was used for dyes – but that is why you should ask its forgiveness before swinging the axe. It is the first tree to colonize swampy land and bogs. Commune with this tree and it will help you to see clearly and free your mind from sorrow and grief. There was a time when we knew all this.

Strangely, there is very little folklore and magic associated with the beech tree, despite its size and prevalence. Perhaps this is because it arrived in Britain relatively late and didn't spread very far, at least not until we gave it a helping hand. It has more resonance in mainland Europe than it does here. The word 'beech' is Germanic and comes from the Anglo-Saxon word for a book, 'boc' – meaning learning or wisdom. And as Danu Forest tells us in *Celtic Tree Magic: Ogham Lore and Druid Mysteries*, 'water from the hollow of a beech tree is excellent for spellwork, blessing, healing skin complaints, as well as gaining the goddess's beauty. As a vibrational essence, beech is used to increase tolerance, compassion, and gentleness.' I've always loved the way water gathers in the shallow, serpentine roots of beech trees. It looks so dark and cool to drink, although I've never been thirsty or brave enough to try any. Despite their extreme beauty (in the autumn especially, when their leaves fan and fall in sheets of gold), and despite the wonder of their colossal grey limbs, there is something cold about a wood that is dominated by beech trees, except on the very hottest of days. I think that's when I'll cast my spells and drink deep of the water from the hollow of the trees' roots.

The branches of the birch, our gentle lady of the woods, were used to make cradles to keep babies safe from witchcraft, and were bundled into newlyweds' beds to boost their fertility. But witches also used them for the brush of their brooms – and you can see the tangled, knotty bunches in the branches of most older birch trees, something that is in sober reality caused by the attacks of the fly agaric fungus. The birch is a beautiful, fertile, welcoming tree, as Dafydd ap Gwilym explained long ago:

> The happy birch wood is a good place to wait for my day-bright girl; a place of quick paths, green tracks of lovely colour, with a veil of shining leaves on the fine boughs; a sheltered place for my gold-clad lady, a lawful place for the thrush on the tree, a lovely place on the hillside, a place of green treetops, a place for two in spite of the cuckold's wrath; a concealing veil for a girl and her lover, full of fame is the greenwood; a place where the slender gentle girl, my love, will come to the leafy house made by God the Father.
>
> FROM *A CELTIC MISCELLANY*

Since those happy days, the birch has been put to grimmer use. William Turner nailed it, painfully, in his 'Herbal' written in 1551. 'I have not red of any virtue it hath in physic; howbeit, it serveth for many good uses, and for none better than for betynge of stubborn boys, that either llye or will not learn.' There were always plenty of beatings of stubborn boys at my long-ago, faraway school, although the birch had been phased out before I arrived. It was indeed a strange place, profoundly old-fashioned even in the 1970s. All the maps on the walls were coloured Empire Red and we learned about Clive's victories over the French in India (and the horrors of the Black Hole of Calcutta) as though they were yesterday. But, as if to compensate, the school was surrounded by deep, magical woods, thick with oak, and ash, and thorn, and these woods and their hidden pathways were wondrous places, where we were left to roam free. There were also the remnants of great elms, and spreading cedars, and beech, and right at the heart of these fairy-tale woods there was a thicket of birch and rhododendrons, clustered around an open space, forming a natural grove. A holy grove! And it was here, we all heard, that a boy not much older than us had once tied up a much smaller boy, stripped him naked and bound him to a tree, and beat him with holly and birch branches until he bled and screamed for help, alone and lonely and far from home. The woods are not always easy and friendly places, we know that, and we do not always walk safely down the 'green tracks of lovely colour'. There is fear in the woods; and savagery; and wild men haunt the holy groves. If some of the ancient mysteries have now fled the woods that may not be a bad thing.

> Farewell to the Downs and the Marshes,
> And the Weald and the Forest known
> Before there were Very Many People
> And the Old Gods had gone!

KIPLING, *FROM* 'VERY MANY PEOPLE'

Perhaps it's fine that the old gods, and more besides, are now lost, despite the last wickerings of the neo-pagans. And yet… we once lived in a world that was alive with gods and spirits and wild possibilities and we have replaced it all with a cost–benefit analysis and a distressed shrug. I don't believe it's the loss of the gods, or God, that matters here. Or I don't want to. But our abandonment of the idea that the natural world can be sacred, and strange, and more surprising or terrifying than anything we might imagine, has cut us adrift from everything that is ultimately important. At the very least, if you believe that a tree has a soul, or that there are powerful gods that dwell in the groves, then you are not going to reduce it all to woodchip. You would not be indifferent to the destruction of the forests. Perhaps that's our real problem: we're just not that fussed any more. We have brighter, shinier toys to play with. 'Nature', for most of us, is something we save for the weekends, or for the TV survivalists and adrenaline junkies, or for slightly weird people. I think what I want to say is that, when it comes to our dealings with the woods, we have lost any real sense of awe.

In the first century AD, when the human world was much younger, the Roman poet Lucan described the Druids and their woods with all the civilized superiority he could muster, but he was also careful not to mock the strange mysteries he was describing:

> ... Hard by there was a grove
> Which through long centuries no hand of man
> Had dared to violate... No sylvan nymphs
> Here found a home, nor Pan, but savage rites
> And barbarous worship, altars horrible
> On crimsoned basements; with the blood of man
> Each tree was sanctified. If faith be given
> To ancient myth, no fowl has ever dared
> To rest upon those branches, and no beast
> Has made his lair beneath... Thus do men
> Dread most the god unknown.

FROM *PHARSALIA III*, 417–22

And when the Romans finally cornered and overwhelmed the Britons and their Druids on the island of Anglesey, and slaughtered these strange wild men and women, they made sure they also uprooted and burned every last sacred tree. Just as it says we must in Deuteronomy, chapter 12, verse 2: 'Ye shall utterly destroy all the places wherein the nations which ye shall possess served their gods, upon the high mountains, and upon the hills, and under every green tree.' Where does this hatred of the woods and trees come from? Who in their right mind would want to chainsaw the Glastonbury Thorn or reduce Sherwood to a splintered, hollowed-out pocket? Is it just because they can? Surely not – or not often. But this destruction is not mindless and it's not greed (or not always). The fact is, we are scared witless by what might be lurking at the heart of the woods – the faeries and the outlaws, the wild untamed green men. As ever, the dirty little secret at the heart of the violence is fear – the primal human need to subjugate anything strange or unknown. It's the boot in the face and the bulldozer at the gates. That's what drives our need to wipe the whole slate clean.

H ere, again, is the Roman, Lucan, writing about the beliefs of the Druids:

> If what ye sing be true, the shades of men
> Seek not the dismal homes of Erebus
> Or death's pale kingdom; but the breath of life
> Still rules these bodies in another age –
> Life on this hand and that, and death in between.

FROM *PHARSALIA I*, 454–8

If what they say is true, it is not just goddesses, magic and elves that are all around us: it is the dead, returned as living bodies, and in this copse, inhabiting or in fact reanimated as the flowers, the birds, the people and the trees. Back down the path towards Rudyard Kipling's old home, there is a small wooden sign to mark the death of his son, John, placed here exactly one hundred years after he was killed in the First World War:

27TH SEPTEMBER 2015

NINE BRAVE MEN FROM OUR VILLAGE AND SURROUNDING AREAS,

INCLUDING JOHN KIPLING, WERE LOST DURING THE

BATTLE OF LOOS IN SEPTEMBER 1915

100 YEARS ON, WE ARE COMMEMORATING THEIR LIVES

BY PLANTING NINE FRUIT TREES IN THE MILL ORCHARD

LEST WE FORGET

RK

'Lest we forget'. Ironically, I'd forgotten that line was Kipling's, although he was writing about God and not man. People plant trees to celebrate new life and to commemorate the loss of it. Whole woods have been created to mark the lives lost in our wars. There were thirty-three woods planted around Britain two hundred years after Nelson's victory at the Battle of Trafalgar; the vast, decaying sweet (Spanish) chestnuts that moulder on the hill at Croft Ambrey were apparently placed there four hundred years ago in the exact formation of the Armada, before it was broken and burned; and there's a grove of Canadian maple trees at the side of the A3 in Hampshire that celebrates their countrymen's sacrifice in the Second World War. If you wander the fields of the Somme, the orderly British cemeteries are all fringed with trees, but the only German cemetery I visited was laid out *among* the trees in the heart of a pine wood. The trees and the stones were jumbled together, and the ground was thick with dead pine needles. The Germans have always thought of themselves as a forest-dwelling people, but what struck me more than this, or the air of distant neglect under the silent pines, was that so many of the memorial stones in this sacrificial grove were emblazoned with the Star of David.

Kipling was devastated by the death of his only son. He'd managed to help get John a commission in the Irish Guards (even though his eyesight was poor), but John only lasted a couple of days on the front line before his face was ripped off by a shell. If he is, as the Druids would have it, 'reincarnate elsewhere', then he couldn't hope for a better place than among these fecund young fruit trees in the bosom of the ancient Sussex hills. He is in the company of eight other local men who died, as Rudyard wrote in rage and grief, 'because our fathers lied'; but John's body was never found and of course none of us really know where he is, or even what he is – other than a bit of Flanders mud. But perhaps ('O Best Beloved') he *is* here, in some form, living out his perpetual existence, and even now he 'goes out to the Wet Wild Woods or up the Wet Wild Trees or on the Wet Wild Roofs, waving his wild tail and walking by his wild lone'.

It interests me, this idea that the spirit of a person lingers in a place long after they are gone. You can feel them in their homes, soon after they've died (or after they've left – we don't have to kill them off…), although you could say that what we're sensing is just the fading memory of someone sitting in a favourite chair, or leaning against the table they once spent so much time laying and clearing. The discarded photograph only slowly bleaches to white. But anyone who has ever visited the fields and trenches of the Somme has felt the loss and desolation in the air. So much trauma and death, they say, has seeped into the landscape that the texture of the world has been changed. In Simon Schama's extraordinary book *Landscape and Memory*, he suggests that it is 'our shaping perception that makes the difference between raw matter and landscape'. In other words, it's us that make the difference, and it's our culture-bound minds that shape what we see and feel in the world – although, as Schama roams through the Polish forests where his ancestors once worked as loggers, he does leave behind a little sliver of doubt.

I have less rigour – or more credulity – which is why I'm standing at the head of an obscure wooded valley in north Devon, not far from the village of Porlock, trying to pick up the ghostly presence of Samuel Taylor Coleridge. It was here – or it was probably here – in a farmhouse just above this wood, that Coleridge fell into a drugged sleep after a long day's walking, and woke to find that he had a poem fully formed in his mind, just waiting to be poured onto the page. The poem – the fragment of a poem – was 'Kubla Khan' and there would have been even more of it – it would, I am sure, have answered every question we have ever had about life, death, art, love and nature – but just as Coleridge was poised to reveal the secrets of the world 'a person on business from Porlock' came knocking, and Coleridge lingered too long at the door, and when he rushed back to his room to finish it, the poem had evaporated. Or so he tells us. And it's certainly a more original excuse than 'the dog ate it, Miss'. But imagine being the owners of this lonely farmhouse, just above –

> that deep romantic chasm which slanted
> Down the green hill athwart a cedarn cover!
> A savage place! As holy and enchanted
> As e'er beneath a waning moon was haunted
> By woman wailing for her demon-lover!

> COLERIDGE, *FROM* 'KUBLA KHAN'

Imagine having this man turn up at your home one night, exhausted but *raving* about demon lovers and the waning moon, high on opiates, crashing late and then lurching from his bed to answer *your* door, scaring the life out of your cousin from Porlock, scattering fragments of genius from his torn notebooks. Imagine trying to tell him, gently, that they're not 'cedars' in the woods, but 'woaks, Sir, *woaks*'. You'd be pleased to see the back of him, although for several months, through the years 1797 and 1798, Coleridge haunted these lonely woods, hills and slippery coastal paths. He walked for miles, for days, unable to settle at home (which was twenty-five miles from here in Nether Stowey); restlessly seeking out his neighbour, Wordsworth, and shaking him with a thrashy torrent of ideas and poetry; plunging through 'wood and dale' and 'forests ancient as the hills'. 'Kubla Khan' is an explosion; it's about creativity, or sex, or what it means to have bipolar disorder – we don't know, except that it contains wild truths. And Coleridge, like Kipling, understood that all true magic must come in threes:

> Could I revive within me
> Her symphony and song,
> To such a deep delight 'twould win me,
> That with music loud and long,
> I would build that dome in air,
> That sunny dome! those caves of ice!
> And all who heard should see them there,
> And all should cry, Beware! Beware!
> His flashing eyes, his floating hair!
> Weave a circle round him thrice,
> And close your eyes with holy dread
> For he on honey-dew hath fed,
> And drunk the milk of Paradise.

COLERIDGE, *FROM* 'KUBLA KHAN'

Well, we've heard it before, but it still packs a punch – and even more so if you are standing at the head of the valley, looking down at the wooded coombe just below the isolated farmhouse, where Coleridge first conjured this magic. It is May Day, traditionally the first day of spring, when the sun returns to a frost-ravaged land, and if I were a young maiden I should be wading through the dew at the foot of the hawthorn tree that is blossoming fretfully to my left, alone in a field of hungry young sheep. At my back is a dark line of pine trees (what else?), looming over the valley and being slapped around by a strong wind, but ahead and down the steep hill to the coombe is a more ancient land, a quieter spot, with a mass of broadleaf trees hazed in their first outpourings of green and, beyond them, a gratifyingly sunless sea.

The first day of spring is always a hard date to agree. Is it the vernal equinox, in most years falling on 21 March? Or is it, as the old tradition had it, on 1 May? Our ancestors lived in colder times, when the River Thames would freeze and the winters were bleak. To complicate matters, what is now 1 May was, until 1752, 19 April; and what is now 11 May was the old May Day. This is when the British calendar 'lost' eleven days, when the 'Julian' calendar was replaced by the new-fangled 'Gregorian' one, and there were riots in the fields and the churches. Mrs J.H. Philpot in her 1897 book *The Sacred Tree or The Tree, Religion and Myth* has this story about the changing calendar and its effect on an offshoot of the Glastonbury Thorn that had survived in Quainton, Buckinghamshire:

> [It] suddenly sprang into fame again when the new style was introduced into the Calendar in 1752, and the people, resenting the loss of their eleven days, appealed from the decision of their rulers to the higher wisdom of the mirac-ulous tree. According to the *Gentlemen's Magazine* for 1753,

about two thousand people on the night of 24th December 1752 came with lanthorns and candles to view the thorn-tree, 'which was remembered (this year only) to be a slip from the Glastonbury thorn.' As the tree remained bare the people agreed that 25th December, N.S., could not be the true Christmas-day, and refused to celebrate it as such. Their excitement was intensified when on 5th January the tree was found to be in full bloom, and to pacify them the authorities were driven to decree that the old Christmas-day should be celebrated as well as the new.

These days, the levels of consumption required to feed two Christmases every year would probably spell the end of the planet, but I mention Mrs Philpot's exciting story because it doesn't feel quite like spring yet, here on the hill above Coleridge's coombe, with only that lonely hawthorn and a straggle of gorse in bloom (and when is the gorse ever not?).

The edges of woods are not simple places and it is sometimes not easy to pass from the open land into the close, skyless company of trees. I am walking down a flinty path, flanked by ragged hedge-rows and curious lambs, with the sun now tentatively shining on the valley. Maybe it's this sudden soft bath of sunshine, but there is an invisible barrier between the sunny fields and the dark wood, and it does take something – not courage, exactly, but a conscious effort – to step from the light into the shade. Once through the gate, though, I am home in the trees' familiar embrace. Or, as John Clare would say:

And this old gate that claps against the tree
The entrance of spring's paradise should be.

FROM 'WOOD PICTURES IN SPRING'

It is right, I think, to pause and lean on a gate at the edge of a wood, before passing through. In any case, there is a man coming up the woodland path, twisting through the trees, and just as he reaches me a cuckoo calls from higher up the valley, the first I've heard for years. 'That was nice, wasn't it?' says the man, his face hidden under a broad-brimmed hat, 'a cuckoo on 1st May.'

May Day should be a day of magic. The cuckoo is a sign of a happy marriage, or of imminent adultery, although it is hard to see how it can be both. Always carry elder twigs... or sew them into your lover's pockets. The cuckoo's calls follow me down into the coombe, past light drifts of bluebells, fat young clumps of nettle and crowds of low-growing holly bushes, now fading back into the woods with the greening of spring. There are violets by the side of the path, their soft lilac faces marked by 'honey guides', the pale white tracks that have evolved to steer insects into their pollen-rich hearts. They're rather like a runway's landing lights, set up to bring the aircraft safely home; and I'm thinking that this coombe, with its infallible path, could be my own personal honey guide, drawing me in, looking for something out of the ordinary. Honey-dew, perhaps. That would... well, that would make it all worthwhile.

I pass a very grand holly tree, growing wild and jagged around its battered old trunk. I can hear the river now as it hastens towards the sea, and then I can see it, a tight-runnelled, restless stream, hustling past bracken and moss-drenched rocks, throwing up sprays of icy light. Coleridge must have walked this way, not so very long ago, and watched the river leap and tumble. And he will have known this oak tree, its great trunk and branches hung about with spring ferns, its young, lime-green leaves tinged with a fading red. There's a tiny, sunken church here, in a tenuous clearing in the woods, and I sit and watch the river race by. The sea is very close, although it is quiet and hidden from view. There are no cars, no people, just birdsong – and sunlight and lichen mottling the ancient church walls. There are sycamores all around, but I am thinking of lime trees, and their slow retreat from the woods, and of Coleridge writing in his prison bower, and of the time I came walking over the South Downs, scrambling down wild rabbit paths, through overgrown woods of ash and chestnut, and then, dropping down the banks of a dizzying gulley, I slipped and sprawled into the last remnants of a lime tree copse, about ten immense trees hidden in fountains of green from the grip of the modern world. They cannot have been coppiced or cut for centuries. They must have been here when the Saxons carved their farms from the Sussex Weald, or even earlier, when the Romans drove the British tribes from their hilltops and forests and marched them into slavery. I kneel and crane to look up at their scoured trunks, their fragile summer leaves, the beech trees all around, crowding in, and then, under a half-fallen elder tree, pushing through last year's leaves, I find a very young lime sapling. It is heartbreaking, the sight of this slender thread with its five green leaves and blood-red buds, hiding in the last refuge of a long-vanished forest. I don't know why, but staring at this sapling, with the holy warmth of these lost limes at my back, fills me with grief and joy.

In fact, I think there's only one thing I do know, as I sit in the shadow of Coleridge, waiting for magic to emerge from the woods on this first day in May. If you go looking, it won't be there.

8

But there is no road through the woods.

RUDYARD KIPLING,
FROM 'THE WAY THROUGH THE WOODS'

Blood in the Forest

If you're looking for genuinely ancient woodland in Britain – and by that I mean not just somewhere that passes the official definition of having been continuously wooded since the year 1600, but is a place that has slumbered, undisturbed by humanity, since the glaciers ground their way back north; and if you tell enough people that you're looking for this wood (and you don't just ask the experts, because they will tell you that 'in all probability' no such place exists, even though it's hard to shake the suspicion that they have squirrelled away some secret haven just for themselves); and if you talk long enough about your desire to find an unspoiled paradise, a wood that will, naturally, over the centuries have been visited by people, and foraged, and hunted clean, but is otherwise, to all intents and purposes, in its original, primordial, Edenic state – then sooner or later someone will always mention Wistman's Wood on Dartmoor. I've put off visiting the place for years, for that very reason. Who wants to have their dreams punctured? But I am here now, on a dreary day in May, with a cold mist dripping from the dead-looking heather, and I am taking my inevitable punishment from Natural England and their wretched, rain-drenched, pissily prosaic noticeboard:

> This woodland was once thought to be relict 'wildwood'.
> However it is more likely that the wood has come and
> gone with the changing fortunes of farming and mining
> with the large granite rocks protecting the developing
> trees from grazing animals. Studies show that the oldest

individual trees are about 400 years old and that the
woodland area has doubled in the last 100 years.

And yet… you will have noticed the qualifying 'more likely'. There
is always hope. We do know that the wood was here in 1632, when
Tristram Risdon described it in his 'Survey of the County of
Devon' and found the trees growing cramped and low, 'no taller
than a man may touch to top with his head'. The existence of the
wood then, replete with dozens of stunted but mature oaks, does
not quite tally with the statement that none of today's trees is older
than four hundred years, although it's possible that the larger trees
have all been felled in the interim. But then, if that were so, we
would still be able to say that the woodland itself has endured.
Perhaps I shouldn't worry so much about this point (although I *do*
so much want to find my elusive ancient wood), because the truly
remarkable thing about Wistman's Wood is not, in fact, its age… but
its trees. They are pedunculate oaks, growing at an altitude higher
than almost any other oaks in Britain, and they have survived by
hiding in an inhospitable rocky gulley, cramping and warping their
own growth in order to avoid the icy storms that roar above – and
it's because of these bleak winds that the trees never grow tall, but
have dipped and twisted and contorted themselves into a dwarf
forest. The sheep are kept away from any saplings by the 'clitter',
the shards of slippery granite that are scattered all around and
have no doubt led to many a sheep's broken leg and slow death as
it strained for a tasty young leaf.

   Risdon and others since may have talked about the trees being
no higher than a man's head, but that is no longer true. The climate
has been warming, the oaks are growing taller, and for the past few
years they have been poking their top branches cautiously above the
wind-blown sides of the gulley. As Natural England notes, the area

covered by Wistman's Wood has also doubled in the last hundred years, something we should all be celebrating, but about which they and the other custodians of Dartmoor seem to be somewhat ambivalent. The problem is that this wood, if allowed to spread, would completely change the moorland scenery, which is one of gorse, heather, grass and sheep. The trees are kept at bay by grazing and 'swaling' – the burning of the moor – and this infuriates many environmentalists, who want to know why we are keeping our vast national park in a self-imposed state of ecological degradation, when native woodland could bring rich and transformative life to the landscape. How can we criticize the Indonesians for fire-bombing their own forests when we are doing the same here? Would we survey the smouldering Sumatran wastelands and insist that they stay that way for ever? Once, long ago, Dartmoor was all trees, before our most distant ancestors turned it to farmland, and then it lapsed into moor. Perhaps it's time to allow back some of the woodland. Natural England's noticeboard informs us that there are just three high-altitude woodland copses on the moor (of which Wistman's is one), but I can stand here at this gate and count half a dozen rigid squares of conifer plantations, standing dark and arid among the heather. It seems strange that their man-made arrival has been tolerated, while the natural spread of the oaks is fought with sheep and flame.

The closest buildings to the wood are in an almost non-existent place called Two Bridges. Just over one hundred years ago, my grandparents spent their honeymoon in the hotel here, in the days before the trees (and the wood) doubled in size. I have a photograph of the newly married couple, my grandfather Edward Montmorency ('Monty') Guilford and my grandmother, Kathleen, both looking so young and smiling so brightly, sitting side by side among the fresh bracken on a bank by the path to the woods. Behind them, just over the close-cropped hill, crouch the invisible trees. If nothing else, it looks like my grandparents were lucky with the weather. I mention Monty and Kathleen, and their honeymoon on Dartmoor in the shadow of Wistman's Wood, because it's surprising (even though I don't think that either of them was especially superstitious and Monty, in fact, had just qualified as an Anglican vicar) – it's still surprising that they should have chosen to spend the first week of their marriage in a place that was known to be haunted by the hounds of hell.

Sir Arthur Conan Doyle knew it. Indeed, any local could have told Monty and Kathleen that the Hound of the Baskervilles, which roamed the moor with blazing eyes and a hunger for human throats, was without doubt a 'Wisht Hound', a member of the ungodly pack that hunted and howled in Wistman's Wood. The pack was whipped on by the Wild Hunt, which all the locals also knew was led by the King of the Faeries, or Odin, or by Satan himself, and was followed by the dead and the damned, thundering by on fear-maddened horses, and that the sight of this hunt foretold disaster (and madness or death for the person unlucky enough to be out on the moor on a moonless night when the Wild Hunt came riding by). And it was Sir Hugo Baskerville, the 'wild, profane and godless man', who brought a curse down upon his family when he flogged out his own pack to chase down a maid who had rejected his violent advances, and who hunted her across the moor, drunkenly spurring his wild-eyed horse, and all the while 'there ran mute behind him such a hound of hell as God forbid should ever be at my heels'. Yes, you must never go near Wistman's Wood after dark, and who cares if Natural England thinks it's possible that the true origin of the wood's name is 'Welshman's Wood' – you only have to see the wood itself to feel the truth of the darker tale.

It is about a mile along the path from the hotel to the wood, and although it is noon (in May) the night does not feel very far away. I can sympathize with Dr Watson and that first, dismal trip to his temporary home on the edge of the moor:

> Suddenly, we looked down into a cuplike depression, patched with stunted oaks and firs which had been twisted and bent by the fury of years of storm. Two high, narrow towers rose over the trees. The driver pointed with his whip.
> 'Baskerville Hall,' said he.

The path to Wistman's Wood follows the river, over rocks and past small gulleys, and it's only when I'm almost on top of the wood that it seems to disengage itself from the mist. Like Watson, I'm looking *down* on the trees – although there are no firs or narrow towers – and it strikes me that this must be a remarkable place in high summer when the trees emerge, at last, from their winter slumber. It's not often that we are able to see a wood's canopy from above, where so much of its attendant animal life gathers. On a hot day in July it should be thick with wasps, caterpillars, moths, beetles and butterflies, along with the birds that feed on them, not to mention the heave and pulse and urgent unfurling of the leaves closest to the energizing sun. But today there are no leaves, just a subtle swelling of the oaks' buds and perhaps a whiff of new life in the damp Devon air.

It's so wet here, the trees look like they're struggling to escape from a surging sea of moss and rocks, although I do wonder if it's possible that they are not growing at all, but inexorably sinking back into the slimy moor and suffocating with desperate slowness, 'as if some malignant hand was tugging [them] down into those obscene depths'. I should have steered clear of *The Hound of the Baskervilles* last night. But the trees do indeed creep sideways up the hill, long branches of mossy oak swerving over the lichen-smeared boulders, while a solitary holly, with a scattering of unseasonal berries, seems to float in the upper branches of one especially Gothic tree. The oaks are no longer head-high. In fact, while the lower trunks and branches look wild and unsettled, the upper branches are growing in a more familiar, oak-like fashion and the trees are now three times the height of an average-sized man. The moss is everywhere, though, great waves of it smothering the smaller rocks and splashing up the lower trunks of the trees. A light wind starts, bringing a cold rain, and the thin sheets of pale lichen that hang from the upper branches flap like torn shrouds in the fretful breeze. Weirdly, and despite everything I've read, there are sheep wandering silently among the trees, nipping at a few sparse tufts of grass. If I look up and over the wood, or behind me towards Longford Tor, there is nothing but bare moorland in every direction, although there seems to be plenty of more succulent grass on offer than the sheep can hope to find among the trees. They're cussed creatures, sheep. And irredeemably stupid.

Of course, they can't be entirely hopeless, the bone-headed sheep, otherwise there wouldn't be so many of them. Most of the animals that humanity has found a use for, the needy and the willing, the pets and the farm animals, have proliferated over the millennia. It's the rest of the animal kingdom, the part of it that wouldn't come quietly, that has been sidelined or exterminated. Especially anything that's bigger than us – we don't seem to like that. Smaller animals are another matter and many still thrive – just look at the bacteria and the headlice – although it may be that there's a plan to subdue them at a later date. As we all should know by now, the reverse is much more likely. For the moment, though, some of the less obtrusive creatures have managed to dodge the centuries of slaughter, and they give the lie to the idea that humanity has won its suicidal War on Nature. All we've done is kill most of the things we could easily find and couldn't tame. The moa and the aurochs. The mammoths and rhinos. The tigers and bears. Some of them (the giant sloths, the dodo, the North American bison) were killed for fun, or out of spite. Is there any point in making a list? Trees, which weren't even able to run away, have been hardest hit of all.

That said, some of the larger wild animals were reprieved... if only so they could be hunted at a later date. In Britain, this has led to the preservation of many of our most glorious woods. A dense wood makes the perfect refuge and nursery for young pheasants, foxes, badgers, hares and every kind of deer, and many landowners decided that they'd rather have the game than the timber. Ever since William the Conqueror laid claim to the New Forest, woods have been conserved and stocked with animals that were kept in the best of health, right up to the moment they were called upon to fulfil their grisly destiny. Not every large mammal benefited. The wolves and the lynxes that once roamed our forests were designated as vermin, with a price on their pelt, and everyone, even the peasants, was encouraged to have a go. The extinction of the wild boar in Britain may have been a miscalculation, although farmers have always hated its rooting and truffling ways. Nowadays, you'll find hunters pleading to be allowed to 'harvest' the boar that have escaped into the woods of Sussex and Somerset – they say that a managed hunting season offers more protection for the species than vengeful local farmers, wielding poison and traps. Perhaps they're right. It's a fashionable theory in parts of Africa, where small-town American dentists and Gulf potentates pay large fortunes to gun down elephants and lions, thereby saving them from being killed at a later date by poachers. If you look at it like that, what boar in his right mind wouldn't prefer to meet his maker under the gentle caress of a high-powered rifle bullet? It's surely better than a painful mauling from a drunk with a shotgun. Although I still don't quite understand why the option of just not killing them at all is not, apparently, an option.

This is not to say that hunting has been an undiluted success for the woods. Most of the Great Caledonian Forest was burned to the ground so that the grouse could prosper in the heather (and to flush out the insurgent peasants); while at the same time, many woods around Britain have been ravaged by rhododendron so that the pheasants (with an 'h') could have somewhere to hide. The hunter and the gamekeeper's loyalty will always be to the animals and the sport, not the landscape, and if that means that whole woods have to be obliterated, or trampled, beaten and dragged, then so be it. But, in general, hunters seem content to let the woods alone, just so long as there continues to be a plentiful stock of birds and beasts ready (if not willing) to be killed for fun. The euphemistically titled Field Sports are never going to be to everyone's taste, but if it's any consolation some of our woods do seem to be thriving amid the bloodshed.

Dr Johnson thought it 'very strange, and very melancholy, that the paucity of human pleasures should persuade us ever to call hunting one of them', but he misses the point. There's a dark glamour to hunting. Perhaps not the kind that involves standing in a damp wood, shivering in line with a posse of overstimulated City traders coming down too fast from their latest infusions of coke – and ready at any minute to loose off two barrels into their own or their neighbour's foot. And there is no logical, or rational, defence of hunting – just a kind of brute bawling that 'man' has a right or a need to sate his primordial bloodlust, or that rural communities are unravelling because city dwellers would rather eat lentils than hot red meat. But there is, I can see, or so I have read, something viscerally thrilling about the chase:

When all the world is young, lad,
And all the trees are green;
And every goose a swan, lad,
And every lass a queen;
Then hey for boot and horse, lad,
And round the world away;
Young blood must have its course, lad,
And every dog his day.

CHARLES KINGSLEY, *FROM*
'WHEN ALL THE WORLD IS YOUNG, LAD'

When all the world was young, and before our technology had outpaced our brains (or at least our capacity for self-restraint), and when the forests and the fields seemed to teem with limitless life, then that may have been the time to indulge our taste for the hunt. From an armchair in my London home, I have gloried in the hunting stories of *The Irish R.M.*, and thrilled to the moment when the master blows on his horn and the riders jostle knee to knee, and the horses leap and lurch over the hedges and scramble down the scree, and the hounds stream ahead across the fields and the drab red fox flits over a wall and streaks for the safety of his distant copse, while the whole hunt whoops and hollers and roars in wild pursuit. Siegfried Sassoon, who hunted obsessively in the aimless years of his twenties before he lost his taste for blood at the Somme, hauls us into the scintillating adrenaline-rush of a day with the hounds; but he also shows how so much of the joy of the hunting is really just the pleasure of being out on a crisp November day, moving easily on a steady horse through a wide and empty landscape, or brushing past low branches, following a winding path in the silence of a lonely copse. The joy of the hunt is entwined with a love of remote nature – not in fact that different from Wordsworth's long walks (although he hated the fact of hunting) – punctuated with short bursts of blood and mayhem.

All of this makes me wonder why a hunt has to finish with a death. Somerville and Ross, who wrote *The Irish R.M.* stories, only show the inevitable end of a hunt once, at which point Mrs Sinclair Yeates bursts into tears at the sight of the dismembered fox. My mother, who loved horses so much when she was a child she thought she was one, remembered her only hunt with horror, when after a glorious day's riding she was cornered by bellowing huntsmen and smeared with the gore of a freshly slaughtered fox. Modern-day enthusiasts, chaffing against the banning of hunting with hounds in 2005, say that it's just not the same, following other hunt members around the country as they drag a heavily scented bundle of rags through the hedges. The hounds, they say, need something to get their teeth into. If that's the case, maybe they could adapt an idea from the Normans, who guarded their hunting grounds with ferocity, and would sometimes sew a convicted poacher into the skin of a stag and set him loose to be hunted down and eviscerated by the pack. Or so the Saxons claimed. And true or not, it would surely be a fitting send-off for any Master of Fox Hounds at the end of a year's hard service, to be tracked and killed by his own well-trained hounds. There's even an ancient precedent. As Ovid tells us in *Metamorphoses*, Actaeon was a consummate hunter who lost his way in the woods after an especially bloody day's hunting and came across the goddess Diana bathing naked in a stream. She was mortified, and then furious, and transformed him into a stag; but even as he retreated back into the forest, not sure where to go, stumbling on unfamiliar hooves, his own hounds caught the scent and he fled for his life, with the dogs howling and ravening at his heels. He tried to cry out, 'it's me, your master, Actaeon', but no sound came other than an anguished wailing, a noise that was somehow neither fully human nor animal, and then, his strength gone, at bay among the rocks, the pack poured onto him:

> And fixed their snouts deep in their master's flesh: tore him to pieces, he whose features only as a stag appeared.—'Tis said Diana's fury raged with none abatement till the torn flesh ceased to live.

OVID, FROM *METAMORPHOSES*, BOOK III

I am probably over-identifying with the victims of hunting, but here, on the edge of Wistman's Wood, with the invisible sun losing its power at the end of a dirty, fog-bound day, I find I am suddenly all alone on the moor, the only sound a distant baying of dogs. It is not a comforting noise. Hunting with dogs has not died away in Britain, whatever the government may hope: it has been driven underground and deeper into the woods. Even in Runnymede, not far from the Ankerwycke Yew, on National Trust land (of all things), there are signs reminding us that poaching is a crime, illustrated with a photo of a greyhound at full stretch, its jaws drippingly agape, closing in on a wild-eyed rabbit. On the day I found these signs, and as I made my way back past the yew trees and the meadows, I saw in the next field a pack of greyhounds, over ten of them, with several terriers yapping and scampering ahead. They were being followed by a group of about five men and women, shouting and whistling to their dogs: it was a timeless scene, something you'd have witnessed if you'd stood here in Henry VIII's day, or much earlier, or at any time until very recently, and you wouldn't have thought anything of it. But now, with just a low hedge separating me from this active, hungry pack, and the image of the traumatized rabbit fresh in my mind, I want to know what's going on. What, in fact, they are hunting. The scene makes me uneasy, and it's not because I disapprove, or because it seems alien and strange. The people could not look more ordinary, like they've wandered out of the local pub. What's most troubling is the marauding power of these dogs and the speed and intensity with which they are casting and circling the ground. Fear is never far away from the fields and the woods – and this is what's really bothering me. I am deeply, primordially, atavistically afraid.

Even so, there can also be something comfortingly old-fashioned about an organized hunt. Every year, in winter time, at the school I attended from the age of eight, a beagling hunt would gather on the 'carriage sweep' in front of the main school building, and the boys would be released after breakfast to mingle with the horses and the local men and women in all their hunting finery. Not only that: for one day only, we were also considered part of the hunt, allowed to 'run with the hounds' in our gym shoes and kit, although it was always clear we weren't completely welcome because the teachers would wade in among us on these snappish winter mornings, kneeing the beagles to one side, and barking at us to keep away from the back hooves ('horses can kick') and avoid the front end ('horses can bite'), and for God's sake stop bothering the huntsmen and women. But we swarmed among them anyway, fidgeting at the horses' harness, breathing on the sparkling stirrups and the black shiny boots, gazing up in awe at these giants in their creaking saddles as they sipped their exotic sherries and smoked lordly cigarettes. And then the master would blow his horn and the beagles would rush and surge, and they'd drain their drinks and stub out the last cigarette (the people, not the beagles), and they'd be off, trotting down the path to the ha-ha, round the side, through a gate to the field, and then disappearing over a ridge never to be seen again. Oh, we ran after them, hoping to see a rabbit caught and shredded, but they were far too quick for us and part of their game must have been to put as much distance, as fast as they could, between themselves and the earnestly shrieking boys. Years later, this joyous escape from the timetabled drudgery of school and into the frosty sunlit uplands of the Buckinghamshire countryside has become tangled up in my mind with the dark glamour of hunting and its promise of unregulated, frontierless freedom. It is a siren call.

> Awake! arise! and come away!
> To the wild woods and the plains...
>
> SHELLEY, *FROM* 'THE INVITATION'

To be sure, the hunt has, on the whole, been good for the woods. We have more of them than we would otherwise – and they may even be a little bit wilder because of the shooters and the riders and their designated game. In fact, there are woods around Britain that lie unvisited by people for months, guarded by gamekeepers, full to bursting with rare flowers and fat pheasants. But despite this, and despite the feeling that hunting offers us a connection to a younger, earthier, less sanitized Britain, there isn't really any excuse. I still don't see how anyone can get their kicks from killing animals for fun; sport should not involve inflicting pain on other beings; and let's never forget that the biggest cock strutting in the woods is always the one holding the gun.

When you think of the woods, does your heart lift and sing? Do you long to be standing at the edge of a wood, among the soft grasses and the smell of ripe wheat, and do you step eagerly onto the path that curves and narrows and disappears into the trees? Are you impatient to discover what waits around the next corner – and do you not care where the path may lead, so long as it takes you deeper and further into the forest? Are you happy to make this walk by yourself? And when you've travelled for hours, and chosen your paths, or maybe you thought you'd strike out and leave the unfamiliar way, and you're now standing alone in the close green silence, are you sure that this absolute solitude was what you wanted? You are, after all, a long way from home. And the night is drawing in.

Well, I can only speak for myself,

> But far more ancient and dark
> The Combe looks since they killed the badger there,
> Dug him out and gave him to the hounds.

EDWARD THOMAS, *FROM* 'THE COMBE'

Bad things happen in woods. There is the hunting, yes, and the dog fights, and the digging up of badgers in their setts. But more than that… people hide their worst crimes deep in the woods, thinking that no one will ever know or discover what they have done, or that the gods cannot see through the black canopy of the trees. On the news, a dog walker stumbles across a body scratched into a ditch, in 'a patch of woodland just outside the town'. There are mass graves in the forests of Poland and Rwanda. Do not go walking alone in the Blue Mountains of New South Wales. Never rent a cabin or go camping with your college friends in the isolated forests of Appalachia. The woods are dark. We do not know where the path leads or where the forest ends. Just think of poor Snow White, taken by the Huntsman deep into the forest, so that he can cut her throat with his sharp-edged knife in a remote glade far from any hope or help. And even when he repents, and lets her go, and she flees further and further into the forest, the trees fight her every inch of the way, lashing and clawing at her face, their roots snaking and tearing at her limbs, and thousands of eyes glare with hunger and hatred from the darkness, and the whole of nature turns with shocking malignancy on this terrified young girl. At least, that's how it goes in the film.

The first time I was ever lost in a wood – properly lost – I was about six years old. I was with my brother, two years older than me, and another pair of brothers, roughly our ages. We had wandered away from our mothers and their picnic and had taken only about three or four turnings on the narrow forest path before trying to head back. I remember standing and talking in a small clearing, faced with a confusion of possible ways. Ignoring everything we had ever been told, we decided to leave the paths and head back by the most direct route. We clambered over fallen trees and pushed through thick, whippy undergrowth, but it very soon became clear that we must be heading in the wrong direction. We started to shout for help, softly at first, but it was summer and our voices were smothered in a density of leaves. We shouted louder. Shockingly, my brother's friend, an older boy, started to cry – and then wail and keen with great tear-choked sobs. He was inconsolable. I did wonder if he knew something the rest of us didn't, but what I remember feeling most, as we stood and suffocated in this ocean of green, with no hope of ever finding our way out, was an unexpected current of interest: we were *children* and we had *left the path* and we were *lost in the woods*. We knew how these stories went. Little Red Riding Hood. Hansel and Gretel. Rumpelstiltskin. Round the next corner there is a cottage, innocent in a sunlit glade; a strange little man is sitting on a tree stump, posing impossible riddles; a sleeping giant stirs and the earth cracks and opens; a hedge of thorns swallows us up; the frog and the fox talk; there is a tower, and a castle, and a sheet of golden hair; the wolf and the witch are on the path and they're at the door, 'come in, my pretties'; and two shallow graves, the soil freshly turned, wait for us behind an old oak tree. Things happen in woods that are outside the runnels of ordinary life, although we were rescued, in the end, by our golden labrador, barging through the brambles with a stupid grin on its face. It was a moment of pure Disney – not Grimm.

Here's another story. Once there were two young children whose parents were dying. So the father called them to him and said: 'I want you to go and live with my brother, your uncle. He will look after you, and guard your wealth until you come of age.' The dying man was very rich. But his brother had a damaged soul, and although he took the children into his home, before a year had passed he was gripped with a desire to get his hands on their money. So he hired two men to take the children into a nearby wood and kill them – far from the sight of God or man. It was wintertime, and there was snow on the ground and no leaves on the trees, and the men led the children into the dark heart of the forest. But when they got there, and he turned to face the children, one of the men realized he could not bring himself to murder a child. Instead, he killed the other man and fled, telling the children he would return with help. But of course he never did. And the children wandered alone and lost in the forest until, in the biting cold, they lay down in each other's arms, and died under an old oak tree. A robin gathered fallen leaves and carefully mounded them on top of the frozen bodies and there they lie, in their shallow graves, for all eternity. Their uncle spent their money, his own sons were lost at sea, his crops failed, and he died a bitter and a bankrupt man. He rots in Hell.

The original ballad of 'The Babes in the Wood' was published in 1595 under the (spoiler alert!) cumbersome title, 'The Norfolk gent his will and Testament and howe he Commytted the keepinge of his Children to his owne brother whoe delte moste wickedly with them and howe God plagued him for it'. The story is not a fairy tale, but a local legend: thrillingly, it seems as though there really was a wicked uncle, a man called Robert de Grey, and the wood where the babes breathed their last, Wayland Wood in Norfolk, has survived – and although it is a fraction of the size it was in the sixteenth century, you wouldn't have to be completely clueless to get lost in it today. Just young and scared. The old oak where the babes were supposed to have been buried was destroyed when it was struck by lightning in 1879. We don't know if anyone thought to dig down to see if there were two tiny skeletons embraced in its roots, but people did come from miles around to gather souvenir splinters from the shattered tree. It's a lovely example of a tree helping us to touch our past.

I am visiting Wayland Wood because it seems like a good place to try to understand my own fear of woods. I am more easily scared these days, and that bothers me. When I was younger I would walk for hours, alone, through the comforting woods of Kent, but I don't feel so at home any longer. Perhaps I'm out of the habit, but it's not just me. The lives we have chosen are prising us apart from the natural world, and we are more likely to experience a woodland through watching *Countryfile* than by breathing in the actual, living trees. Most of us are cloistered far from any woods, and the ancient connections have faded. Almost no one works in the forests any more, and so we imagine we don't really need them. In fact, the great impulse of the past few centuries has been to build ourselves a more controllable world where the woods and the wild no longer matter – one where we can live behind walls, flicking switches, and the good things in life will be delivered to us down wires and pipes and through the front door. As the disconnect grows, so does the fear. And so it goes. The staggering success of the low-budget 1999 film *Blair Witch Project*, in which a group of film-makers find themselves being slaughtered one by one in the woods, and dragged away into the darkness, all of it filmed on close-angled, bewildering, smothering home video, with no hope of seeing what is lurking among the trees, or even just behind the person who is filming – well, that really did tap into our deepest fears.

Anyway, that's why I'm heading for Wayland Wood (or 'Wailing Wood' as it was once known), passing signs for Wayland Prison (home at one time to Reggie Kray and Jeffrey Archer) and the Wayland Freemasons Lodge. For some reason the presence of these two institutions adds to the atmosphere of Victorian Gothic that I've managed to cultivate in the car by listening to M.R. James's ghost stories for three long hours. When I was away at school in the 1970s, during the three-day week, whenever there was a power cut in the winter (and, oh, how we prayed for them), the teachers would have to suspend normal lessons, light candles and read us stories to keep us quiet. Amazingly, rather than choosing something soothing like *Winnie the Pooh* or *The Wind in the Willows*, their most popular choices were Saki (especially the vicious little tale that ends with a tyrannical aunt having her throat ripped out by a polecat) and M.R. James. There's a particularly chilling story called 'Oh, Whistle and I'll Come to You, My Lad', set in a Norfolk seaside hotel, not far from here, in which a man finds a medieval whistle and, rather than flinging it into the sea, takes a good long blow – with the predictable result that he's followed back down the beach (and ultimately into his hotel room and then his bed) by a shapeless human form that is blindly, hungrily searching for something. Or someone. They read us that story more often than any other.

It's the end of a long day in early May and the tiny car park is gratifyingly empty. The gate to the wood is open and the ride is wide and welcoming.

> Shudder all the haunted roods,
> All the eyeballs under hoods

Shroud you in their glare.
Enter these enchanted woods,
You who dare.

GEORGE MEREDITH, *FROM*
'THE WOODS OF WESTERMAIN'

The trouble with my plan to try to scare myself (half) to death in Wayland Wood is that the wood itself does not want to play along. It is a beautiful day, there's a languorous droning of insects, a wood pigeon is thrumming out its tune, and the soft late afternoon sun is right now coaxing the bird cherry blossom into one final foaming ecstasy of display. There are frothy bursts of bird cherry at the entrance to the wood, their flowers achingly white against the green of the leaves, and there's also one particular tree, about sixty feet high, standing alone in a recently coppiced clearing, that is shaking the air with elemental vigour and pungency. The whole wood seems high on its perfume – and in no time at all, so am I.

There are flowers everywhere, some of which I can even identify. On the edge of the coppicing there are primroses and sprays of forget-me-nots, purple thistles, bluebells and creeping red campion. Young nettles are tentatively flowering in beds of young dock leaves (I was once told you could prove the existence of God because He so thoughtfully placed the dock next to the nettle) and there are buttercups nodding in and out of the sunlight, beaming and dimming, like the flashing of dozens of miniature lighthouses. The coppiced hazels have mostly shed their yellow catkins by now, but the tiniest nubs of nut have taken their place, gathering for the summer's growth, and the hazel leaves are already thick on the twigs, and darkening. Everything, all around me, is new and fresh and ragingly green. Even the ash trees, always the last of all, have spent their flowers and are on the very brink of leafing.

I have a very sketchy and curtailed knowledge of wild flowers. My mother and my aunt loved them, and knew many of their names, but when I was a child I was more interested in climbing trees, or using a stick as a bazooka, and when I was a teenager it was quite obvious that neither of them knew anything about *anything*, and then, somehow, the moment passed and I left home. So it isn't long before I find myself staring at something mauve and wishing they were here with me now. Instead I have a 1947 edition of *Wild Flowers of the Wayside and Woodland*, compiled by T.H. Scott and W.J. Stokoe (based, it says, on the standard work *Wayside and Woodland Blossoms* by Edward Step), which for some reason I'd grabbed on my way out, rather than something more modern and, frankly, accessible. Anyway, it seems I am looking at a bugle (*Ajuga reptans*), 'common in spring-time in wood and field, and on the waste places by the roadside'. It was once used by herbalists to treat bruises and ulcers. I wish I'd known that, although moments later I find I am able to identify a snake's head fritillary (*Fritillaria meleagris*) without Stokoe and Scott's help, a lovely, drooping, purple and pink flower, whose torn-edged, bell-like flowers do indeed look like sleepy snakes' heads. Apparently there are adders in these woods, although surely it's too early in the year, or late in the day, for them to bother anyone.

I have never known so many flowers, of such variety, to be crammed into such a small wood. Perhaps I never looked before. There's a beautiful little light-mauve flower, teetering on a long stem, that is almost certainly, probably, a germander speedwell (*Veronica chamaedrys*), once also known as the forget-me-not until it was superseded by the more familiar, light-blue garden plant. The other name for the flower is 'cat's eye' but, given the velvety softness of its four perfect petals, maybe cat's paw would be a better name. All of a sudden I am on my knees in Wayland Wood, snuffling at tubers and stamens, an instant expert on wild flowers, vaingloriously renaming and mislabelling everything in sight. I decide I have found *Lamium album* (that's a white deadnettle, obviously – although according to Scott and Stokoe it is so like a real nettle that it 'cannot be distinguished at a casual glance'). My glance is not casual. The deadnettle gets its name from the Old English 'deffe-nettil', meaning powerless. It has no sting, although I'm not so entirely confident in my new botanical skills that I want to actually grasp the thing.

This is a very well-ordered wood. The coppicing has been done on strict and regular rotation and there are neat stacks of sawn timber piled at the edges of the clearings. There's a chainsaw muttering somewhere in the distance. Just off the path there is a small spread of iridescent bluebells under light tree cover and, standing up proudly among them, there are six early purple orchids (*Orchis mascula*), each with about twenty fluorescent flowers hanging like birds in flight from their stout purple stems. It's Britain's most common orchid (and the first to flower), but despite its beauty I'm afraid it smells of cats. If you dig under the vigorously swelling stem you'll find two testicle-shaped tubers – for very obvious reasons the orchid has always been considered a mightily potent aphrodisiac, although this vital local knowledge has faded of late. Maybe we should consider offering plates of orchid tubers to advocates of Chinese traditional medicine in exchange for their rhino horns, before they kill the lot. Or maybe that would just lead to the extinction of the orchid as well as the rhino. Male insecurity is endlessly destructive. What, in heaven's name, is wrong with Viagra?

Leaving the sunlit clearings for the dark wood is not so easy. There are fewer flowers, although there's a great bank of cow parsley (*Anthriscus sylvestris*) glittering to one side of the narrowing path. In truth, I'm starting to doubt Scott and Stokoe's handiwork: according to them, there is no such thing as cow parsley (although I've been bandying the name around for years), and I may in fact be looking at something called 'cow parsnip' (*Heracleum sphondylium*), or 'hogweed', named after the Greek hero Heracles because of the strength of its dead hollow stems. Are they making this up? I wish I knew. But I am lost to the world in a heady new enchantment of wild flower names: earthnut; nipplewort; nodding bur marigold; mouse-ear chickweed; butterbur; bristly oxtongue; false Cyperus; bastard balm; treacle mustard; virgin's-bower; wake-robin; stinking goosefoot; blunt-leaved hawk's beard; penny-pies and water pepper. The ploughman's spikenard. Woad! The zigzag clover. Can I have my life again, please? I'd like to pay attention this time.

I wander among the flowers, and the wood closes in. There's a pond to one side, ringed by blackthorn and willow, its waters looking black in the sunshine. I look for life on its surface – maybe water skaters or tadpoles – and at once I remember the other ghost story that one of the more sadistic (or clueless) teachers used to read to us by candlelight when the electricity failed on those long winter nights. It was about a village taxidermist, who liked to gather frogs at night from the ponds, and stuff them, and dress them in little costumes and arrange them in tableaux in his shop window for the amusement of his customers: weddings, games of cricket, musical concerts, that kind of thing. One evening he went out as usual, to gather more victims for his business, but the wind was blowing hard that night, and the trees seemed to press close, and when he stood gazing at the pond, net in hand, he could see the eyes of thousands of frogs staring right back at him. He turned to leave, but he felt ivy coiling around his feet and branches feeling for his arms... and the next morning he was found by the side of the pond, stitched and stuffed, a look of terror on his contorted face. At least, that's how I remember it.

It's a powerful idea, that one day Nature will take her revenge on those who abuse her, and it's one we cling to today, with our

Gaia theory and the hope that something will rise up and put an end to the senseless, endless human hunger for MORE. William the Conqueror's son, Rufus, was killed by an arrow when out hunting in the New Forest, but everyone knew it was because he'd upended the Natural Law, and dispossessed the people of their land, and killed too many and too much. And the weird blue creatures in the film *Avatar*, with their Native American drumming rituals and deep connection to the Tree of Life – they are rescued by Mother Nature (and a man with a gun) from the technology-crazed mining corporation that wants to blow up the forest and dig out every last nugget of mineral wealth. It's a great, violent, green victory against the ideology that says that everything has a value to be extracted and exploited. Although, sadly, it is only a film, and we may have to do more than just clap in the cinema. I realize now, perhaps like many other people, that I've been subconsciously waiting for the pendulum to reach the end of its swing – and that I have always assumed there will be a moment when the burning and looting of the planet will reach its furthest point and then the earth will come boomeranging back in a blaze of green glory. But what if there is no such point? What if it's not a pendulum at all and there are no natural checks and balances or Great Greek Goddesses? It seems more likely, doesn't it, that we're the only ones holding on – and the rope's no longer attached to anything – and we've just launched ourselves over a cliff.

The coppices on either side of the path are older here, and it's impossible to see far into the interior. On one side there are bluebells glimmering in the fading light, but the wood to my left is turning cold and dark. We didn't always walk at ease in the woods, and I can see why travellers in Chaucer's day would band together before making their journey. If something did come out of the trees, running low through the sun and the branches, you'd be half-blinded; it would be onto you before you could move or even shout. It's important to stay on the path, we know that, but there is a faint track disappearing past a weeping bird cherry and when I push at the branches, swiping aside showers of white blossom, I see a broken tree, torn off at the base, and all around it there are the most beautiful white butterfly orchids. Or perhaps they are stars-of-Bethlehem (starflowers). The light is too faint to read the faded illustrations in Scott and Stokoe. It's an enchanted spot, and the last of the track (such as it is) leads me on to more orchids, and bluebells, and sprays of honeysuckle, and even a wild service tree (the infallible sign of an ancient wood), with six shabby, coppiced trunks hauling themselves to the sky. On and in. There's a two-stemmed hornbeam, its bark flashing embers of pink and grey in the last of the light. A wood pigeon is purring. I've never seen so many bluebells. Ahead, there's another wild service tree, but this one is cramped and its trunk is twisted into something like the shape of a minotaur's head. The wood pigeon stops its drone and, instead, I hear that the chainsaw has started up again. There are cars in the distance, but it's an alienating and distant sound in this murk. It's amazing how quickly a wood's mood – my mood – can flip. All I can say is: never leave the path as night is falling; and never look behind you: once you start, you won't be able to stop. All of a sudden I have become convinced that someone is following me, moving easily through the trees and the shadows, keeping just out of sight. There is absolutely no reason why they should, but it is a long time since I've been alone in a wood in the twilight. I do wish that chainsaw would stop. The wood pigeon bursts out of the trees behind with a violent clatter. I have succeeded in making myself scared, and I cannot now remember why I wanted to: all I know is that I want to get to my car, fast, but I mustn't run because that'll bring the person – the thing – after me and he – it – knows these

dark woods better than me. So I walk, hunched and alone, and
the darkness thickens, and I don't look round, and my whole back
flinches and cringes in anticipation of the heavy footsteps and the
crushing impact.

> Like one, that on a lonely road
> Doth walk in fear and dread,
> And having once turn'd round, walks on
> And turns no more his head:
> Because he knows, a frightful fiend
> Doth close behind him tread.
>
> COLERIDGE, *FROM*
> 'THE RIME OF THE ANCIENT MARINER'

It's ridiculous, of course, I know that now – and Oliver Rackham,
the pre-eminent (and twinkly, but also somewhat testy) woodland
writer, who loved these particular woods, would be scathing: I am
surrounded by wonders – the bats are darting through the treetops –
and all I can do is fuss about being massacred by a Norfolk maniac
with a chainsaw. I have been away from the woods too long.

The only time I experienced real terror in a wood was in Queen's
Wood in Highgate, a patch of ancient woodland in north London.
It's not a big wood, and in most parts you can see the houses looming through the trees, but late one autumn evening I was shuffling
through the leaves, admiring the hornbeams, the drifting ivy and
the flapping strands of police incident tape, when I saw a man
with his dog, approaching through the dying light of the day. At
that exact moment, everything seemed to go quiet, other than the
hiss of the autumn wind in the trees above, and as we passed one
another, the man took a too-sudden step towards me and said,
'It's best not to walk here after dark', then hurried away. It gave

me a shock. I was in the heart of the wood, at the bottom of a small dell, in a place where several paths converge. The wind was picking up and it was later and darker than I'd realized, and when I looked up there was a young woman standing at the far end of one of the paths, dressed in a drab white dress, holding a baby. She didn't have a buggy, or a sling, she was just holding the baby, which seemed strange – somehow old-fashioned – and she was staring at me. I think I half-waved, in an attempt to make things normal, but then, in the half-light, and with my eyes on her, she vanished, or rather she *faded* into the gloom, leaving me alone among the trees. And immediately, very close, a dog barked hysterically, and dozens more dogs, from every direction, began to howl and rage. I was shaken by a great surge of adrenaline – and this time I didn't stop to think that it was ridiculous, or that Oliver Rackham would disapprove. I just ran.

It's possible that we have cleared and destroyed our woods because we fear their wildness, in the same way that when we're scared of the dark we want to throw open the curtains or switch on all the lights. John Fowles thinks that people have two conflicting souls, a 'green' one and a 'dark' one, and that in most people the dark soul has won, which is why we are scorching the planet with such apparent hatred and venom. But my hope is that we've just forgotten what it's like to live among the trees – and if we spent more time in the woods, we would be comforted and calmed, not scared, and we'd remember to treat them well. In any case, it's such a waste of time worrying about imaginary ghosts and axe men lurking in the forest. There's a storm coming and there's so much more we should be doing to get ready. We're about to find we need the woods more than ever.

9

Outside,
A world in sunshine;
She with her hand in mine:
Such a wide, dark flood;
I died in it, where I stood –
By the side of Nellie.

ELIZA KEARY,
FROM 'THROUGH THE WOOD'

A River Runs Through It

The Hebden Beck starts slow and cold. Pushing out of the reservoirs of Walshaw and Widdop, it gathers its tributaries from the moors south of Haworth, where Heathcliff once roamed, before picking up speed as the land falls towards the woods around Hardcastle Crags. By the time the river has left the open moors and has reached the current upper limits of the woodland, it's fast and dangerous – but also useful, as Abraham Gibson found when he set up his cotton mill here in the year 1800.

There are no fewer than twelve named woods in the two-mile ribbon of woodland that unfurls from the head of the valley to the point where the Hebden surges into the outskirts of the Yorkshire town of Hebden Bridge. People tend to lump together all these copses and plantations under the general catch-all 'Hardcastle Crags', but it seems a shame to lose the names. They speak of a patchwork of local ownership, of generations of stewardship and exploitation, whereas now they've all been snapped up and rendered one by the National Trust. But the names linger on, and although we wouldn't miss the prosaic Lee, Ingham and Shackleton Woods, I think I'd like to know what prompted the naming of 'Foul Scout Wood'. Was there one especially snotty-nosed boy who was forced to pitch his tent alone in this patch of woodland, while the rest of the pack built their fires and polished their woggles just across the valley in 'Dill Scout's Wood'? Was he unkempt? Unwashed? Possessed? How long did he have to live in his little tent under the trees before the wood was named in his honour?

You can always tell when a place is owned by the National Trust. It's not just that everything is tidy and spruce – woods as well as buildings – there's also something particular about the people who visit and the way they act once they're in the Great Hall at Hever or on the coastal path near Bude. You could say it's because everyone's wearing sensible shoes, brandishing a map and drinking from a thermos flask, but it's not just that. It's also because of the way we behave.

Most of the time, people hesitate on the edges of woods; there's always, at the very least, a flicker of a moment when we pause and gather ourselves before we pass from the light to the dark. It may be that primal reflex, dating back to the times when the woods held predators, but I realize now it also comes down to a question of ownership. Most people assume that, to them, most woods are out of bounds. And yet, as I stand among the Scots pines in the National Trust car park at the southern end of Hardcastle Crags, I watch countless people stride into the woods without a backward glance. It probably helps that it's a bleak day in mid-February and there's a soft but insistent damp chill settling from the moors, but there's no doubt that these people feel right about being here. They feel proprietorially at home. The road is good and they are soon sheltered and muffled by the trees.

The river defines these woods. It has scoured out the sides of the valley, on which the trees cling and occasionally slip, but it also fills the air with damp and noise. Every trunk, branch, twig and last, stubborn brown leaf is slick with moisture, a mist is rising from the river and the ground and the leaves are heavy and greasy on the forest floor. Scrambling down a narrow path brings you to the river's edge, but it would in any case be impossible to get lost in these woods. The noise of the river, in its mid-winter flood, is overpowering. Further upstream, around Gibson's Mill, the water has been penned and trammelled, but here it hurls itself through the valley, surging over the weir and foaming against its banks. The water is a rich, malt brown, glossy with peat from the moors. It looks, somehow, good enough to eat.

I am amazed, once again, by the resilience of trees. A spindly willow is growing out of a backwater and, at the river's fiercest point, an ash tree has fallen, its roots in the air but also still holding fast to the bank. The tree's trunk has bridged the river and it has thrown at least a dozen branches skywards, all of them decorated with a light-green filigree lichen. Mid-stream, there's no competition for light from other trees and the ash looks solid and powerful and a sturdy match for the pounding waters. The Vikings venerated ash trees for the speed of their growth and their power. The World Tree, Yggdrasil, which grew from the underworld to the heavens and spread its branches over the earth, was supposedly an ash tree. A massive serpent called Nidhogg coiled around its roots and an eagle roosted in its topmost branches, while a squirrel darted up and down the trunk, spreading gossip and taking messages between the two. Harts and serpents (and even, by some accounts, a goat) devoured its leaves and branches and stripped its bark, but it was replenished by the waters from the spring of Uror, sprinkled on its aching leaves by the three maids of Norn. Indeed it was.

> The ash Yggdrasil
> Endures more pain
> Than men perceive,
> The hart devours it from above
> And the sides of it decay,
> Nidhogg is gnawing from below.

The Tree of Life, in other words, existed in a perpetual churn of growth and decay, pain, loss and creation. Odin, the King of the Gods, is said to have hung himself from this tree for nine days and nine nights, where he had one of his eyes pecked out by ravens, gaining an inner eye, or wisdom, in exchange. He was presumably happy with the deal, and the pagan Norsemen who worshipped him spread their own sacred trees across their lands; or so we believe, because they left almost no trace of their world other than their decaying boats, looted treasure and hundreds of millions of silver coins, exchanged with traders from the East for the traumatized peoples they captured and sold as slaves. At the most sacred tree of all, on the site of a massive temple in Uppsala in Sweden, the eleventh-century writer Adam of Bremen tells us that every ninth year the worshippers would hang nine male heads of every living creature from its branches: dogs, horses, sheep, rats, men... their blood would drench the leaves and placate and soothe the gods. Or so said the Christian Adam, eager to spread word of the utter barbarism of his pagan ancestors. But it's interesting, isn't it, that the Vikings, who worshipped the Tree of Life and believed that every tree had its own spirit, and that this spirit flowed through every living creature, were also the most notorious butchers, rapists, murderers and dealers in staggering quantities of enslaved men, women and children – despite what the revisionists may claim. Hugging a tree is not an automatic path to love, enlightenment, peace and harmony, although further to the East, the Buddha's sacred bodhi tree does not carry quite the same stench of slaughter.

Anyway, it's likely that the Vikings most valued the ash tree for its bountiful production of long, straight, supple shafts for their spears and bows. The ash pollards well, although if you are thinking of making yourself a spear, or perhaps a longbow, you should cut it before the sap starts rising in the spring. It'll last longer – and fly truer. Left to their own devices, unpollarded or coppiced, ash trees don't grow straight. Their young growth might, but after a while they have a tendency to meander. Branches droop and it's only the black-budded tips that curve upwards. The Vikings may have made use of the ash for war, but in Britain people believed that the ash had the power to heal. They spooned its sap into their babies to make them strong, scraped and drank its bark to cure constipation, and passed their weak and ailing children through a cleft in its trunk to infuse them with the strength of this deep-rooted tree. There is more magic associated with the ash than any other tree. It was even said by Pliny that waving a full-leaved branch in the air would ward off snakes; or, in John Gerard's *Herball, or Generall Historie of Plants* (1597), that 'if a serpent be set between a fire and ash leaves he will flee into the fire'.

Now, of course, the ash tree is in the news for all the wrong reasons. In 2012, panic gripped the land because the media and, to its eternal shame, the government, had at last noticed the arrival of ash dieback disease, the grim and seemingly unstoppable fungus *Chalara fraxinea* that was spreading through British woods, from the East, and killing most of the ash trees in its way. It is depressingly easy to find numerous references to ash dieback in books and articles from long before the government decided to announce its presence, and there's no doubt the authorities had a good chance to try and stop a disease that is now beyond their control, but almost nothing was done until it was already too late. Ron Freethy writes about ash dieback in *Woodlands of Britain* in 1986; that's about twenty-five years (*twenty-five years*) before the government took hold of the stable door and rather resentfully pushed it half-closed. As Freethy perkily notes, in blessed ignorance of the impending two decades of blinkered and ideological stalling: 'the cause has not yet been positively identified although the Department of Forestry at Oxford

University are currently researching the problem'. I wonder how they got on.

Oliver Rackham is very clear that the root cause of ash dieback is globalization: the unfettered free movement of goods, services, people and, disastrously, plants and trees around the world. There are more restrictions now, but they are still inadequate and reactive: as with ash dieback, by the time the government's bureaucracy has ground into action, the damage is usually done. The opening-up of borders to non-native trees and other plants has had dire and bitter consequences for almost every corner of the planet and its local ecologies. In Britain, Rackham tells us in his last, short book *The Ash Tree*, 'about as many introduced tree diseases have appeared since the 1970s as in all the years before'. As the drumbeat of world trade quickens, governments have always seemed keener on barking about some imaginary red tape, or cossetting the insatiable needs of big business, than they have on protecting our native habitats. Until very recently, you could buy invasive species on Amazon and eBay. Perhaps you still can. In the last one hundred years the losses have been harrowing, all across the globe, brought on by plant diseases and insects inadvertently (or even deliberately) introduced into vulnerable populations, from the pines and oaks of Kyoto, to the red oaks of the United States, the eucalyptus of south-west Australia, the pine trees of Crete, the plane trees of France, European cypress trees, the alders, box, horse chestnuts and larches of Britain, and now our oaks (threatened by acute oak decline) – it's a desperate, heart-wringing list. Whole forests are disappearing – and we always react too late. Rackham even gives us the answer: we should ban the import of trees commercially. Simple as that. Small numbers are fine, but only so they can be properly checked before they're allowed in. What has happened instead is that ash dieback disease has been helped on its way by fresh plantings of infected, imported ash trees, snuggled into the soil across every part of Britain. We'll never catch it now.

I've taken to checking the ash trees I meet for signs of ash dieback, although I'm not even entirely sure what I'm looking for: black spots on the leaves, leaf wilt, black strips on the bark, death... Right now, there are more ash trees in the UK than there are people. Imagine if we lost them: the emptiness that would open up in our woods, hedgerows, fields, streets and parks; the void in our hearts. I am just old enough to remember the desolation of Dutch elm disease in this country, but not old enough to have understood the anguish. And anyone aged under fifty would not know or feel what we lost then. Some of the elms have suckered and are growing again, although the beetle that brought us Dutch elm disease is still prowling the countryside, settling on any elm shoot that dares show itself above the hedgerows, like a low-flying drone taking out a succession of startled meerkats.

Rackham believed that the ash tree would prove resilient to dieback. Maybe we'd lose several million trees, but some would survive and recover. His book was published just in time to acknowledge the next threat to the ash: the emerald ash borer, a green beetle from East Asia that has already destroyed about forty million ash trees in the United States and is likely to account for the rest – all of them – even though the American authorities are now releasing millions of Chinese wasps to eat the beetle's larvae and eggs. It's like the old woman in the song who swallows a fly, and then a spider (to eat the fly), and then a bird (to eat the spider) and so on – until at last she swallows the horse (she's dead, of course). We could follow this arms race of pests and their predators, with who knows what consequences, or we could learn to nurture and cherish what we have.

Humanity and the ash tree have lived together in Britain ever since the last Ice Age. About five thousand years ago, when there was the first great wave of elm deaths, the ash moved quickly to fill the gaps. It has been spreading ever since, given another boost in the 1970s when most of the remaining elms were killed. Now it looks like it's the ash's turn, because if the dieback and the beetle don't get it, then, tragically, it turns out that our booming populations of deer will: ash saplings are their favourite foodstuff. We are going to miss the ash. As well as spears and bows, the toughness and elasticity of ash has made it perfect for any kind of handle (axes, garden forks, spades), oars, shafts, crutches, 'the felloes that make up wheel rims' (or so says H.L. Edlin), ladders, pulley blocks, snooker cues, cabinet making, and so on. Its ability to absorb shocks makes it perfect for frames (the first vans and buses all used ash), but it also bends well. The wings of the Second World War Mosquito aircraft were made from ash. Its timber is beautiful: creamy white and wonderfully supple, smooth yet resilient. Here's John Evelyn:

> In short, so useful and profitable is this tree that every prudent Lord of the manor should employ one acre of ground with Ash or Acorns to every twenty acres of other Land: since in as many years it would be worth more than the land itself.

The ash is one of nature's friendliest trees – its Latin name, *Fraxinus excelsior*, is a shout of joy and wonder – and if I linger over the ash, it's only because I'm going to miss it. It's an astonishingly lovely tree – the Venus of the Forest – and so happy to please. Nicholas Culpeper tells us that a decoction of young ash leaves in white wine 'helpeth to break the [gall] stone and expel it, and cure the jaundice', but it also serves to 'abate the greatness of those that are too gross or fat'. And if you place an ash leaf on a wart and recite (perhaps nine times) 'ashen tree, ashen tree, pray buy these warts on me', you can expect them to disappear; although if you live in Cheshire you are more likely to rub the warts with bacon, cut a notch in an ash tree and put the bacon in the hole – in time, the warts will leave your hand and appear as discoloured bumps on the tree's bark. A bunch of ash leaves in your home will protect you from witchcraft; newborn babies must always be blessed with a bowl of ash leaves in water; and all farmers should know that they only need to bury a shrew in an ash tree for the twigs and leaves taken from that tree to cure their sick cattle. As I say, there is more magic associated with the ash than any other tree. It is the best of luck to find an ash leaf with an even number of leaflets springing from its stem: there are usually nine or eleven, one at the top then four or five pairs lower down, but if you can find one with an even number, then put it in your glove:

> The even ash-leaf in my glove
> The first I meet shall be my love.

I don't know why we're not doing more to try to save our ash trees, although Rackham is no doubt right when he says it's too late. But after the flurry of panic in 2012, most people have moved on and are gloomily accepting what feels to be beyond our powers of influence. The Forestry Commission is urging us to wash our feet (and our children's and dogs') after every walk, but if you suspect that not everyone is following this advice, then the battle is already lost. The ash was always considered a calm, thoughtful and rather modest tree: of all the trees in the woods, its leaves are invariably the last to arrive, and it's devastating to think that we've done so little to help. The famous, perky little rhyme is misleading – I've never seen ash leaves appearing before the oak's – but it may also soon be redundant:

Oak before ash, in for a splash.
Ash before oak, in for a soak.

Although on this damp February afternoon near Hebden Bridge there are of course no leaves of any kind on display, unless we count a few sprigs of preternaturally early woodbine twirling around a sodden sycamore stump.

It doesn't matter, though, because at any time of year, and despite my grim brooding over the ash, this little patch of woodland down by the weir in Hardcastle Crags is one of those elusive places that bathes and soothes the spirit. A magical place, seemingly like a thousand other places across Britain – a wood, a river, a track – but somehow there's more. The path turns a bend, following the river, and comes to a cluster of Scots pine trees. It's dry here – or relatively so – with a blanket of needles on the forest floor. There should be red squirrels. Or deer. Maybe it's the calming draughts of pine resin, although the scent is muted in midwinter. The river is calmer and on the far bank there are some plump black birds with white bibs – dippers – hopping on the stones, sipping at the water. A young silver birch tree perches on the river's edge, its greedy sprays of twigs filling the sky. You'll want to stay here for ever.

Round the next corner, nailed to a beech tree, the National Trust has left a familiar notice:

ALTHOUGH THEY MAY LOOK NATURAL, MOST OF THE
TREES YOU SEE HERE AT HARDCASTLE CRAGS WERE
PLANTED IN THE 1870S. BEECH, SYCAMORE AND

SCOTS PINE WERE PLANTED, NONE OF WHICH ARE
ACTUALLY NATIVE TO WEST YORKSHIRE.

BY SELECTIVELY FELLING SMALL AREAS OF TREES,
PARTICULARLY CONIFERS, WE WILL GRADUALLY
RESTORE THE WOODLAND TO A MORE NATURAL STATE.

OVER TIME THIS WORK WILL IMPROVE THE STRUCTURE OF
THE WOODLAND, HELPING CREATE BETTER CONDITIONS
BOTH FOR WILDLIFE AND FOR OUR VISITORS.

Or, as Gerard Manley Hopkins would have put it, had he been in charge of sign-writing at the National Trust:

> What would the world be, once bereft
> Of wet and wildness? Let them be left,
> Oh let them be left, wildness and wet;
> Long live the weeds and the wilderness yet.

HOPKINS, *FROM* 'INVERSNAID'

As we know, people whose task it is to protect the wildness and wet are more passionately concerned than ever with what is 'natural' and 'native'. The words are used three times in five sentences in the National Trust's notice – and it shows how far we have come since the days when the guardians of Britain's forests thought it was acceptable to poison (yes, *poison*) the native trees and cover the land with conifers. Now the conifers are to be rooted out. And not just them, but also the non-local beech and the foreign sycamore trees. I had been admiring one of these renegade beech trees earlier. It had looked much older than its 140 years – the trunk was far wider than my embrace and its thick, slippery roots had taken up more than half of the path – but now it seems as though its days are numbered, along with all the other interlopers. I wonder, what will be planted in their place? Oak, of course, it gets every-where; but also maple, wild cherry, hawthorn and blackthorn. Will they now bother with ash? Perhaps wild service trees, so that five hundred years from now someone will triumphantly identify an ancient forest. Oliver Rackham can get rather impatient with the new tree planters, who lazily replicate the same species mix wherever they go. He says they need to be much more sensitive to the local ecology. But of course it depends how 'natural' the National Trust wants to be. They could take the end of the last Ice Age as their jumping-off point, and then it would be an unbroken stretch of hazel and lime, with beaver, lynx and elk thrown in for good measure. It's easier to nail a notice to a tree than to agree on what is 'natural', especially after thousands of years of human intervention. But, on reflection, I am profoundly grateful that they are trying. We all want to get back to Eden. Even if we have to build it ourselves.

Natural or not, the woods around Hardcastle Crags have been copiously colonized by something called the hairy northern wood ant and I am keen to find these creatures. February, of course, is not the best time to be ant hunting – they should be hibernating, or sheltering underground from the Yorkshire chill – but these ants are hairy and, more importantly, *northern*, so I don't see why they shouldn't be out and about. They're not southerners and it's not snowing, so I am sure they'll be scurrying about industriously, pro-tected by their thick winter fur. In fact, I have a very clear vision of the hairy northern wood ant, sashaying single file along the forest floor, their long glossy coats swinging from side to side like a pack of well-groomed, miniature Afghan hounds.

Their nests, when I find one, are huge – easily over one metre tall – a suggestively hairy, shaped dome of pine needles, twigs and forest debris. There is no sign of any ants, although I stare at the crevices for many long minutes, waiting for one to appear. I don't prod the nest with a stick because I'm aware that this ant, as well as being northern, can spray a nasty jet of ammonium if provoked, and I fancy it would be quite easy to get its hairy back up. Consulting my book, I am disappointed to discover that the only part of *Formica lugubris* that is remotely hirsute is its *eyebrows*. In fact, given its surly Latin name, I now see that there is no chance of me seeing a hairy northern wood ant in winter. It'll be holed up in its empty den, bleakly contemplating the awfulness of life through the bottom of a bottle and lugubriously twitching a pair of shaggy Russian brows. The furthest south you'll find one of these wood ants is in the forests of southern Wales – they like it cold and damp – and I wonder what will happen as the climate disintegrates and drought and storms become the norm. Perhaps they'll adapt. And moult. Or migrate. Most likely they're one of the thousands of species that are facing extinction worldwide. But as I turn for home, I can't help but feel a twinge of irritation: if they're not going to come out and show themselves on what is, frankly, a really quite mild February day, then *what is the point of them being so hairy?*

Darkness arrives quickly in February. There is a brief moment when everything turns orange, the rocks and ferns, the pink-tinged sycamore bark and the last golden leaves of the beech, but then the colour leeches out of the wood. The pathway back down, which had seemed enchanted on the way up, is fading and uncertain and I am suddenly aware of the loud urgency of the river's call. The path offers choices that hadn't seemed to be there on the way in and there are forks and trails that must be invisible when travelling in the other direction. It is always too easy to get lost in the woods. The dark shapes in the undergrowth are rhododendrons, whose days at Hardcastle Crags are surely numbered, and then, looming out of the murk, a massive grey larch blocks the way. It's the ugliest of winter trees, a conifer that drops its needles, leaving a gaunt and jagged profile with clusters of dessicated cones. It seems drained of all life, of any possibility of a return to life, in a way that no other deciduous tree ever does, even in midwinter. Perhaps it really is dead. Sudden oak death (which affects the larch) emerged in South Wales recently, and has led to the felling of over six million trees in an effort to delay its spread. The first part of the disease's name, *Phytophthora ramorum*, literally means plant killer; a variant, *Phytophthora kernoviae*, arrived in 2003 to colonize rhododendrons and beech trees. There's not much in this little paradise that isn't at risk.

Our delayed (and probably ineffectual) response to the mutating diseases of globalization is one self-inflicted problem; but as I follow the throaty gush of the Hebden Beck back into town, it's easy to remember that the greater threat, and the one that is going to amplify everything else, is the looming catastrophe of our collapsing climate: just ask the people of Hebden Bridge, whose town (at the time of writing) has now been flooded three times since 2012, most recently in the winter of 2015–16 when their streets, shops, businesses and homes were deluged by several feet of floodwater and the emergency services fought in vain to find a route through the waters to reach them. There are very specific, local reasons why the floods are multiplying here. This is Yorkshire, after all: the rains fall and the rivers run fast. There will be floods. But the lovely, dark, rich colour of the Hebden Beck? That comes from the peat that is being washed off the moors in unprecedented quantities, because there's almost nothing left to hold it in place. The area around Walshaw Moor is a blanket bog of extraordinary global rarity: and much of it is right here, on the moors above Hardcastle Crags. But the landscape is not as nature evolved. There should be trees and shrubs in the gulleys and drifting and dotted across the moors, their roots stabilizing the soil. Instead we have unleashed the sheep – and they have cropped the place bare. Worse, we have dug channels to drain the moor and we are now burning it every year to keep the heather low so that the grouse – yes, the grouse – can flourish (but only until the day they are driven towards the blazing shotguns on the Glorious Twelfth). The town of Hebden Bridge is flooding, and the moor is drying and dying, to give pleasure to a bunch of entitled country house fantasists.

Of course, that's far from the full story. It rains a lot here – and as the climate warps, and much of the rest of the planet dries, we are going to thank our lucky stars that it does. But the area needs better flood defences (walls, channels, banks) and the moor needs more trees – just as it needs to be left free to perform its natural function as a massive sponge, free from burning and drainage. It has become commonplace to point out that most flood defences are being built in the towns, while almost no effort is being put into averting the floods further up the hills, but that is surely the better way. We should plant reeds, shrubs and woodland; create ponds, gravel pits and lakes for run-off. Drive back the sheep and the grouse. Release the beavers! They'll do the job for free. It's not like any of this is surprising. Humanity has known about deforestation and soil erosion for thousands of years – indeed, Plato looked out at the mountains above Athens, which were once covered with trees, wildlife and flowers, and were now arid, infertile rocks, scoured clean of topsoil, and wrote:

> In comparison of what then was, there are remaining only the bones of the wasted body… all the richer and softer part of the soil having fallen away, and the mere skeleton of the land left behind.

He could be writing that today, standing on a hillside in China or Europe, India or Indonesia, watching the trees tumble and the soil slip and slurry from the land. But there's still time to save the hills above Hebden Bridge.

Again, there's more. The blanket bog is not just a water sponge, it's a carbon sink: it absorbs and holds huge quantities of carbon. When the moor dries or is burned, that carbon is released into the atmosphere. This is terrible news for our efforts to stem the effects of what most people are still calling 'climate change' or 'global warming', although both terms seem far too gentle (even rather pleasing) for something that is more likely to resemble the Ten Plagues of Egypt. As I write, not far from here the county council of North Yorkshire has given the go-ahead to the extraction company Third Energy to frack for shale gas in Ryedale near the village of Kirby Misperton. That decision is going to directly affect the amount of flooding we can expect here in Hebden Bridge – and all over the planet. Every climate scientist in the world would agree (well, ninety-seven per cent of them and counting). It is unfair to blame the put-upon local councillors, even if they did receive over 4,375 objections to the scheme and only 36 in favour. There is huge pressure from our governments, and their friends, all of them desperate for fracking to get under way: there's gas and large quantities of short-term cash to be extracted and funnelled out of the area. Apart from the fact that the locals would probably prefer not to live in an earthquake or cancer zone, nor for hot flames to pour from their tap next time they run a bath, we all need to agree to leave fossil fuels in the ground. If we burn them, much of the world will flood. Other parts will scorch and die. We know this. We'd rather not know (it's so exhaustingly responsible and grown up), but we have to learn to say 'no'. Honestly, though, these local NIMBYs with their health-and-safety concerns, and their desire to live in a picture-perfect, fresh-air Yorkshire village. Frack 'em.

The weird thing is, that this fracking decision was taken in the middle of so much good news about renewable sources of energy. The cost of solar panels is plummeting. Offshore wind is booming. Longer-term electricity storage is within reach. More than forty per cent of Denmark's energy comes from wind, while Portugal generates more than fifty per cent of its energy from renewables. The Chinese are now the world's largest producer of solar power. On the day I write this, the Germans have supplied almost all of their domestic electricity needs through renewable energy. This is not a story you'll hear trumpeted in oil-addicted Britain. I suppose the old-time energy companies are scurrying to extract what they can before economic reality catches up with them, or perhaps before they're sued for wilfully poisoning humanity. Who knows, maybe they'll be prosecuted for crimes against humanity at the International Criminal Court. Some of them have known for decades that their actions were destroying our climate and our health, but they carried on regardless. Is that a reasonable definition of 'evil'? To know that your actions are killing millions of people, displacing tens of millions more and polluting whole ecosystems, but you suppress the information and press on regardless? I think it may be.

So instead of fracking, we should be embracing renewables; and we should be planting. The world needs its woods more than it needs a quick fix of gas. We know how useful trees are: they exude oxygen, stabilize the soil and make the rains fall. They provide shade from the sun and shelter from the storms. The vast majority of species on earth, from the jungles to the oceans, rely on trees for nourishment and life itself. I know we do. And as we merrily pump carbon dioxide and other greenhouse gases into the atmosphere, the forests are doing what they can to mitigate the madness, sequestering over one thousand billion tons of carbon in the trees and in the soil around their roots. The woods of Hardcastle Crags shouldn't be an isolated strip in an overgrazed, swaled desert of bog and farmland; they should be part of a great, nationwide patchwork of interconnecting forests. 'Only suicidal morons,' snarled John Fowles, 'in a world already choking to death, would destroy the best natural air-conditioner creation affords.'

Sometimes, it is surprisingly hard to find peace in the woods, but it is always good to try.

*

Books! 'tis a dull and endless strife:
Come, hear the woodland linnet,
How sweet his music! on my life,
There's more of wisdom in it.

WORDSWORTH, *FROM*
'THE TABLES TURNED'

On a close, grey day in mid-June I arrive at the edge of Glover's Wood in Sussex, a couple of miles west of Gatwick Airport. I am eager to lose myself in the trees – no, more than that, I want to hear the sweet music of the woodland linnet and bathe in its native wisdom – but instead I am fretting about the noise of the planes as they rumble and screech and accelerate overhead. Is it because of my age, or is it the age we live in, that it can be so hard to get lost in the woods? If it's any consolation, even William Wordsworth, the greatest of nature writers, could sometimes struggle to feel anything for the natural world:

The world is too much with us; late and soon,

Getting and spending, we lay waste our powers;–

Little we see in Nature that is ours;

We have given our hearts away, a sordid boon!

This Sea that bares her bosom to the moon;

The winds that will be howling at all hours,

And are up-gathered now like sleeping flowers;

For this, for everything, we are out of tune…

WORDSWORTH, *FROM*
'THE WORLD IS TOO MUCH WITH US'

Glover's Wood is genuinely ancient, semi-natural woodland, although all of it has been thoroughly worked, chopped and regrown over many centuries. There are four large ponds in the wood, and this part was once agricultural land – fields and hedgerows and cattle wading in the water – but they are now given over to oak, ash and hazel, dragonflies, mosses and something called the 'elongated sedge'. It's possible that the wood was also part of the first flourishings of industrialization, in Tudor times, when iron ore was smelted in furnaces in the woods with fuel taken from coppiced hornbeam and oak. Running through the wood there's a steep valley, or ghyll, caused by a stream cutting deep into the native limestone, and both English Nature and the Woodland Trust (to which about one-third of Glover's Wood belongs, with the rest in private hands) are sure that the sides of the ghyll are areas of primary woodland – in other words, they have been continuously wooded since Neolithic times, even if they've been considerably mucked about with. It's a lovely place to be, in mid-June, when the hawthorns have bloomed and the creamy blossom has given way to clusters of hard, green berries, and the grasses are high and laden with seed and cuckoo spit. I am sitting on the edge of the ghyll, on a tangy bank of wild garlic, and

I'm staring at a dead, rotting wild service tree that has toppled for-wards into the fork of a coppiced hornbeam, and my dog is noisily eating the grasses with a side-order of garlic (she – we both – may regret that later), but all I can hear is those passing planes, and all I can really feel is a distracted buzz of low-level agitation. The fumes of the car journey are still with me and it is so hard to leave behind the grit and babble of the city.

We are always told that finding a wild service tree is a sign of an ancient wood – and there are plenty of them here, in Glover's Wood, mostly lurking in the hedgerows, but also standing meekly among the more boisterous holly and hornbeams. I've only recently trained myself to recognize one. It's a solitary, spindly tree, rather shaggy around the boughs, with demure bouquets of small white flowers in June and hard little mottled brown fruits in the autumn. These fruits (called checkers) were once added to beer to sweeten its taste, and on their own could even make a destabilizingly strong alcoholic punch, which is probably why you may sometimes find yourself drinking in 'The Checkers' pub, although it's best not to order a pint of their finest *Sorbus torminalis*. John Evelyn, among others, is in no doubt that the fruit is good for the digestion and a safeguard against colic and dysentery, but he is also eager to tell us that water distilled from the flowers and leaves is a helpful cure for the 'green-sickness of virgins'. The timber is hard and long-lasting, useful for spindles and screws, and it's also a fine charcoal tree, but it has never been of much general use because it's so rare. I like its name, 'wild', and its underdog appeal, and I like the way the one I'm staring at now is slumped so comfortably in the arms of the coppiced hornbeam, not sleeping but rotting, its coarse bark prised open by scuttling woodlice and crumbling onto the forest floor.

There's a slow-moving stream at the bottom of the ghyll with a solitary young holly growing on the far side. A robin is balanced on the top, bouncing up and down, ruffling its feathers, looking here and there in its quick-eyed way, dipping at the stream before flurrying back to another branch. There's no one here apart from me, the panting dog, and the busy robin. Two eyes, shaped like something you might see painted on the prow of a Greek fishing boat, have been carved into the trunk of the hornbeam, although when I look closer I can see that it's not a carving but natural growth – an accident of art. Still, I think, that's a lucky sign, surely – and at that moment it's as though the wood has slowed and opened and birdsong emerges at last from the depths of the trees, and also, even louder, from overhead in the thick, close-leaved canopy; and it's then, right then, that I become aware of the heady reek of wild garlic, and damp leaves, and the soft gurgle of the stream, and the insistent peep of the robin (which is now beseeching my attention from a tree stump just three paces away), and there's so much moss everywhere, creeping over the hornbeam and wild service tree, and mounded in deep-green, billowy piles by the side of the stream, smelling rich and loamy, that I have to resist an urge to roll and bury myself in it; but more than that, there are huge tangled haystacks of honeysuckle in full bloom, engulfing a hornbeam sapling, topping a twenty-foot holly tree and looping over the nearby hazel coppicing. I have no idea how long I have been sitting and brooding on this bank, oblivious to the birdsong, the moss and the honeysuckle. It's not as though it has all just suddenly arrived, because everything has been here all along, obviously, and it's me who has been elsewhere – disconnected and out of tune. Although, to be clear, that's not to say that any part of this wondrous woodland scene could actually care less what I think or am feeling.

Right on cue, another plane screeches overhead, tearing at the forest canopy. It's easy to see why so many scientists are now saying that the earth has entered an entirely new geological epoch – the

Anthropocene Age – defined by the fact that the planet's ecosystems and geology have been permanently altered by human activity. There's no place on earth – not even its atmosphere – that hasn't been changed by our relentless expansion; and this wood, that only a moment ago seemed permanent and timeless and teeming with life, now feels unbearably fragile. Measured by body mass, over ninety-five per cent of the world's vertebrates are now either human or dependent on humans for their existence – sheep, cows, chickens and pets – and just five per cent are wild animals, a tiny, dwindling number of tigers, frogs and otters. Even one hundred years ago, you could have flipped that percentage. Humanity (or a small, privileged part of humanity) has set the whole planet to work for its gratification. In fact, if you want to visit a place where the wildlife is flourishing, the best you can do is head for the area around Chernobyl, the site of the world's worst-ever nuclear disaster. With the people gone, the trees, wolves, elks, lynxes, eagles and wild boars have returned in numbers not even seen in the nearby nature reserves. In the absence of farmers, foresters, hunters and car drivers, and our everyday chemicals and pollutants, nature is finding a way back – just so long as we're ready for the Giant Mutant Bison that, ten years from now, will almost certainly be rising with a righteous bellow from the radioactive earth and marching on Moscow.

Our momentum is frightening. Right now, the well-funded operators of Gatwick Airport are lobbying to be allowed to add another runway to its capacity – to make room for all the extra flights they calculate that we will want to be taking. The local YIMBYs are protesting (yes, in my backyard I would like to be able to live without the tiles from my roof being shaken off every seven minutes and for the air I breathe to contain a reasonably non-toxic mix of oxygen and nitrogen, but without too much extra, climate-destroying carbon dioxide, please). It is baffling, that with the world's scientific community united in agreement that our fuel-burning, airport-expanding activities are going to destroy our planet, we are still giving the time of day to people who want to carry on with business as usual. For sure, it's understandable that most of us are daunted by the changes that will be needed, although the majority of them are undeniably positive – fresher air, fewer floods, clean energy, a growth in community, more worthwhile jobs and a shared sense of purpose. More woods! But carrying on as though nothing is happening is easier for most of us (while it's still possible). We need leadership – someone to help us move the box of chocolates to the other end of the table and out of our addicted reach. And we need to remember that the hot air emanating from the climate-change deniers is just that: the outpourings of a bunch of overindulged charlatans, none of them scientists, with a childish resentment of being told what to do by others who are more qualified. We should turn our backs on these shrieking attention-seekers – the newspaper columnists displaying themselves at the bars of their cages, the needy opinion formers sniggering at the chance to say something different, the corrupt ex-ministers suckling at the oily teat of big business, the asset owners and energy companies funding fantasy, made-up science – and get on with the grown-up business of fixing this mess without them.

Spring came early this year. I know that's 'weather' and not 'climate', but there was something especially unsettling about standing in a London garden in mid-January, with the hot sun on my back, daffodils blooming in the flowerbeds and a swarm of fruit flies streaming from the overheated compost. The woods are going to be devastated by these changes and every expert agrees that the trees and their accompanying species need as much time as we can buy them to adapt to what's coming: the earlier springs, the floods and droughts, the erratic winters. And we need to build stepping stones of nature so that every scrap of diversity is utilized. We should fill our cities and towns with trees and cover the land with woods. The first land mammal to become extinct because of man-made climate chaos – the Bramble Cay melomys, an eager little rodent that lived on a tiny island on the Great Barrier Reef – simply ran out of space as the waters rose.

All of this is hard to imagine, as I walk in the green heart of Glover's Wood. The aeroplanes are still rumbling overhead, but it's cheering to come across a decaying Woodland Trust sign, with its perky acorn symbol and reassuring map and succinct summary of its plans for the wood. Their volunteers have been busy coppicing, and widening the rides, opening the canopy and encouraging back wild flowers such as the greater butterfly orchid and the goldilocks buttercup. Rare woodland butterflies are returning (I keep an ignorant eye open for the silver-washed fritillary). They are leaving the fallen trees to rot, so that the invertebrates flourish and the birds feast and thrive. This place is loved and popular. There are benches near the ponds, dedicated to people who cherished the wood: 'In memory of Dr Clare Tarplee,' one reads, '1962–1999, who grew up in the Mole Valley and loved woodland.' And this one, 'dedicated by a loving family to the memory of Dot and Chas Wren who once owned this lovely little piece of England.' Thank you, Mr and Mrs Wren (or Mr and Miss?), your gift has enhanced many lives. Throughout the wood, there are little camps and dens, where children have been playing and grown-ups have been practising or demonstrating their forest skills. The place feels well-worked, and used, as any wood should.

Above all, there's a peace here that feels beyond anything that Gatwick can throw at it. I sit on one of the benches and stare at the green hornbeams and let myself be soothed by the woodland magic. The crumbling remains of a vast coppiced tree are in front of me, the roots covered in moss and soil, old branches and trunks sprawled in every direction: the whole thing looks as though it has risen from the earth, alien, like it shouldn't be here, or like a giant squid flopping its last on the shore. My dog has disappeared, tracking an imaginary squirrel. Leaning over the pond is a hornbeam, its branches hanging low over the sedge, its lizard-like, pinkish trunk covered in little stimpy clusters of twigs and soft-toned leaves. The hornbeam is an unassuming tree, but it's strong, with a hard wood that holds its long thin branches horizontal to the ground. It has a very strokeable trunk, I realize. My dog comes back, another failed hunt completed, and as I sit here, breathing in the wood, people drift by, nodding and friendly, or lost in their thoughts. The Woodland Trust was founded in 1972 by Kenneth Watkins, a retired farmer, who had become angry about the wholesale planting of conifers and the loss of his local copses in south-west England. Since then, the organization has planted over thirty million trees and is guarding over one thousand woods. They are just one of several charities dedicated to the protection and proliferation of native woodland. If you love woods, it is reassuring to remember that you are not alone. And it is also good to know that in the battles ahead to save our woods and ancient trees (and, indeed, much more than that), there are many millions who feel the way you do.

10

One impulse from a vernal wood
May teach you more of man,
Of moral evil and of good,
Than all the sages can.

WILLIAM WORDSWORTH,
FROM 'THE TABLES TURNED'

This Wood Is My Wood

At the time of writing, it's still just about possible to say that Smithy Wood is a fifteen-hectare remnant of ancient woodland, cradled in a gentle curve of the M1 motorway, a short distance north of the centre of Sheffield. There's plenty of evidence that the wood has been here for at least the last eight hundred years, although once a wood is that old you can plausibly trace its lineage back to the last Ice Age. Like any English or Welsh woodland that has been in continuous existence since the year 1600 (the cut-off date is 1750 in Scotland), it is officially designated as 'ancient', for the simple reason that before those dates no one felt the need to plant woodlands: they must have seemed to be endlessly abundant. In its pomp, Smithy Wood would have been part of a great expanse of northern forest and you could have strolled from here to Sherwood without once leaving the shade of the canopy; but over the centuries the wood (or what's left of it) has been parcelled off, gnawed, nibbled and uprooted, its trees used for housing, heavy industry, the railways, mining and fuel, until, at the end of the 1960s, the M1 was driven through its centre, further splintering the last scraps of woodland. In short, its fate is fairly typical of much of Britain's woods, but at least some of it remains – and the surviving patches live on as part of Sheffield's green belt, the city's enviably lush encircling band of woodland, parks and heath, bringing pleasure to its citizens and oxygen to their lungs.

Ancient woodland is nothing like a newly planted wood. We believe it takes many hundreds of years for woodland to achieve the extraordinary diversity of plants and animal species that together constitute an ancient wood, but in truth, no one really knows how long is needed: we've been too busy chopping down the forests to pay enough attention to the complicated business of rebuilding them. The soil is the key – in an ancient wood it is thick with dormant native seed, invertebrates, fungi (including the mycorrhiza that is symbiotically essential to trees' health) and the accumulated leaf mould and mulch of the ages. It's true that the ancient broadleaf woods that were felled and poisoned by the Forestry Commission and the state-sponsored landowners to make way for conifers have been easier to restore than many thought possible; and the new-planted plantations often failed to take root because the soil was somehow resistant to the drab, septic rows of pine – indeed, it's a romantic and reassuring thought that the earth will reject the imposition of something that is so clearly antagonistic to its needs – but we shouldn't kid ourselves. In most cases the transplant was deemed a success. The longer the conifers remain, their roots purging and transforming the soil, the less chance there is of an easy return to the original mixed broadleaf woods.

Anyway, that's the thing about ancient woodland – every single wood is uniquely adapted to its local circumstances, and once it's gone, you cannot just plant a new one somewhere else, or expect to be able to artificially re-create what was there before. A wood is not just a collection of trees. Of the woodland left in the UK (which has so much less tree cover than almost any other European country), well under twenty per cent is officially 'ancient', and that figure includes something called PAWS – 'Planted Ancient Woodland Sites', where the dominant new planted tree species is often the conifer. This is why the tiny amount of ancient broadleaf woodlands that has somehow survived the centuries of human depredation is unimaginably precious and we should be cherishing every last acre. In fact, it seems incredible that ancient woods are not on a national register and protected by law from destruction, damage or misuse, except in cases of overwhelming national interest. But they're not. Despite their global significance, they're prey to the most fleeting and frivolous planning applications, encouraged by a recent government desperate to put 'economic development' above any other considerations. Of course, it's hard to agree what trumps an ancient wood. Perhaps a new train line (although not if all it's doing is knocking thirty minutes off the journey time from London to Birmingham at a cost of £50 billion); or new housing (once the alternative sites have been thoroughly investigated); and maybe, sometimes, an ancient wood has to be uprooted so we can build some transformative industry, if there really is no other possible site for the project. But local people should have their say, and there should be national laws in place to protect the woods from the people (almost never local) who see a chance to make easy money from dead trees and insufficiently protected land. There is an unaccounted cost to us all (financial, physical and spiritual) every time we lose one of our ancient woods.

I'm visiting Sheffield's Smithy Wood, on the banks of the M1 motorway, because there's a proposal from Extra MSA Group to turn a large part of the wood into a new service station, complete with parking for 639 vehicles, a food court, shops, a hotel and 300 full-time jobs. On the company's website you can examine its picture of a shiny forecourt, sparkling cars and a sweep of sympathetic architecture set against a backdrop of lush green woodland and a blue, cloud-dappled sky. If Extra MSA gets the go-ahead, in mitigation for the loss of a chunk of Smithy Wood it is committed to creating and managing six hundred acres of 'new and existing' woodland as part of 'The Chapeltown Community Woodlands' for the next fifty years. The plans for this wondrous new woodland really do look lovely. It is true, as it says on MSA's website, that 'the creation of Sheffield Services does require the use of around twenty acres of existing Woodland', designated as 'Ancient Woodland... within the Green Belt', but MSA has 'adopted an informed designed approach' – and the pictures of their futuristic woodland look soothing and wholesome and undeniably tidy, with wheelchair users, joggers and happy cartoon children enjoying themselves on a golden, sunlit path at the pruned edge of a wall of green trees, just a doughnut's throw from the new food hall.

My main problem with all of this, it turns out, is actually find-ing Smithy Wood. The nearest road, once you've left the whump and boom of the M1, curls down Smithy Wood Drive, past perky postmodern office units, and into the Smithywood Business Park. There's nowhere to park and no signs for Smithy Wood, just lumpen blocks of grey, red and blue corrugated warehouses, tightly mowed, Tellytubby lawns, a profusion of new plantings (saplings and shrubs and banks of hot red annuals) and many signs warning the visitor that they're on private land. At the far end of the business park the road merges with an older route, heading back into Sheffield, bounded by ancient hedgerows and meandering under an old bridge before passing yet more industrial units, but nowhere are there any signs for Smithy Wood. Is it being eased out of view, so no one will remember that it was ever here, or mourn its loss? Grandfathers will drag their dogs across the glittering tarmac of Sheffield Services, muttering 'There used to be a wood here once, you know.' 'Yes, Grandad, of course there did... and would you like a Krispy Kreme with your tea?'

I am here on the longest day, 21 June, and a single red kite is mooching up and down under a miserable, cloud-clogged sky. The next storm is not far away. I know from the satellite maps that Smithy Wood must be here, close to this business park, but every-thing – the road, the warehouses, the lighting and the trees – is new but also somehow half-finished and disorientating. Trucks grind past. A man with a shock of wild grey hair parks his camper van in a cordoned-off layby ('no parking' it says) and leaps out with a dog, but before I can stop my car he has scuttled off down a path towards a bank of trees. Is that Smithy Wood? I find some parking reserved for visitors to one of the warehouses and hurry after him.

The man has disappeared but the path leads to a woodbank, a mounded wall of earth at the edge of the trees, and one of the key indicators of ancient woodland. Once it would have been topped with a fence or hedge to keep out hungry livestock and to mark the wood's boundaries, but now, like almost every other woodbank, it is covered with trees, grass and wildflowers (and, in this instance, a smattering of rusty beer cans). A path dips into the darkness and I follow. There are sycamores, horse chestnuts, several scarred beech trees and an ancient ash – the only one of these species native to the area. The path is rutted and pocked. Drifts of plastic bags and bottles are pressed into the mud. A dessicated brown conifer has been hung upside down by a rope from the bottom branch of a battered beech tree and an image forces its way into my mind of Mussolini, his beaten dead body slung by its feet from a girder in a town square at the end of the Second World War. Or maybe, given that this is probably a discarded Christmas tree, I should be thinking of the crucified Christ. At one time, the worship of sacred trees morphed easily into the worship of Jesus and his wooden cross – and to this day I find it hard not to touch wood at almost any moment of uncertain fortune. Woods are good places to visit if you suffer from this irritating superstitious tic. I touch the beech tree for luck: it has had a large T-shaped slab of bark carved from almost half of its lower trunk and the dead Christmas tree is swaying and bumping against the wound. It's not a sign of a happy or a playful wood, although many of the other trees have been spattered with paintballs, so at least someone must have been having fun.

After only about a hundred paces I catch sight of traffic lights and a line of idling cars at the other side, and it becomes clear that this is not Smithy Wood, or if it is, it's only an isolated fragment carved out by the business park. I do what I should have done all along, and turn towards the roar of the M1 motorway. There's a deep-rutted path, thick with growth on one side – hazel, mead-owsweet, blackthorn, dog rose and goose grass – and I follow this, with the birdsong erupting from my left and the buzz of a saw and the trundle of a forklift truck coming from the grey and red ware-houses down the hill. If we're going to keep growing (our economy and industry, our population and housing), then we're going to have to get used to this ever-closer mixing of the 'natural' and the human. Of course, endless growth of everything is not necessarily desirable – and certainly not logically possible – but what does an ancient wood have to offer someone who would rather have a job? Isn't this really the only question? Why should anyone in Sheffield care about saving Smithy Wood and its mysterious mycorrhizae, when the alternative is construction work, new infrastructure and future jobs for a hard-pressed community? We might as well go back and live in the trees, or a cave, if we're going to start blocking the road to the future. The ancient woods are going, or gone. Get over it – and plant a new one somewhere else.

At least I've now found Smithy Wood. The path that skirts the edge offers all sorts of different ways into the trees, some of them narrow single tracks, others wide and muddy and ploughed by heavy vehicles. I leave it as long as I can (for some reason I'm especially reluctant to let go of the open skies today, despite the doom-laden clouds), but eventually I duck under a low-hanging hazel and enter the woods. It's dark here; and still. Someone, perhaps from the business park, has been dumping barrels and wires and plastic detritus, and the outer reaches of the wood are fringed with the shabby cast-offs of modern life. The high green bracken and the violet flowers of the arched brambles hide some of it, as does a rising tide of young birch trees, but there's no avoiding the smear of flytipping. Deeper in and there's self-seeded oak, alder and ash. Mostly, though, as the forest path curls towards the sound of the motorway, it is beech that emerges as the dominant woodland species, the trees probably planted as timber over a hundred years ago, once the demand for coppicing had fallen away. The path opens into a grove of widely spaced beech trees and a blackbird shrieks a warning, briefly puncturing the incessant moaning of the motorway, which has been following me around the wood like the background throb of the *Terminator* soundtrack. The ground in this grove, under its beech-soft green canopy, has been churned and mashed by a thousand spinning tyres, ripping up the woodland banks and shredding the trees. The blackbird falls silent. There's something sullen in the air.

The wheel tracks lead onwards to the M1, past half a child's pink plastic toy telephone. The way is slippery from the storm, with oily puddles filling the joyriders' churn, as the path swoops over short, steep hills and plunges into tree-lined gulleys. In all honesty, it must be exhilarating to drive through these woods. There's one last slope and then the motorway appears through the dense summer leaves, its full noise and power revealed, cars and trucks shrieking north through the rain. How many of these drivers, I wonder, would like to see this ancient woodland replaced by a service station? How many of their fuel tanks or sugar levels are so low, or their bladders so full, that they cannot wait a moment longer before they make another stop? Apparently, for safety and convenience, drivers should be able to pull over every thirty minutes on a motorway, and if you're arriving here driving south from the M18 you may have to hold on for a few extra minutes beyond that limit before you can satisfy your needs. Is that what we want and where our priorities lie?

Heading south through this uneasy wood, tracking the low, barred fence that divides the trees from the road, slipping through piles of discarded rubber tubes and plastic jerry cans, the path reaches a more open area of oak wood pasture. The trees can't be more than one hundred years old, and there's enough light drifting through their young branches to leave a floor of long grass and a scattering of buttercups, nettles and red campion. Beyond the fence is a flower-filled meadow. There's no space between the trees for the racers and the path is narrow and steady. A crow, a daub of pure, living black, lifts itself from a fence post and flaps slowly across the fields. By some quirk of geography, the noise of the road has muffled. There are birds singing in the oaks. In fact, as I stand here in the wood, under the dripping trees, despite its battering from the off-roaders, and the rain and the junk, and the roar of the motorway, and the impending appeal from Extra MSA Group who want to see it razed and tidied away, despite all of this the wood is working its familiar magic. Of course it is.

If you're driving on the M1, about seventy miles south of here you'll pass signs for the National Forest. I'd stopped there once, at one of the many small woods – this one almost next to the motor-way – that together make up the two hundred square miles of the National Forest. The forest was started about twenty years ago in a part of England (it's *England's* national forest) where at the time there was only six per cent tree cover. Since then, over eight million trees have been planted and the forest has been spread across almost twenty per cent of the land – they are aiming for thirty per cent and many millions more trees, creating a huge area of woodland (new and ancient), meadows, wood pasture, farmland, towns and villages. I stood in one tiny part of it, Billa Barra Hill, a nature reserve on an old Saxon burial ground, among the rocks and under Scots pines, oaks and sycamores, in a blaze of foxgloves, with the rain sluicing through the sodden trees, and I realized then why I'd felt so confident that day in Glover's Wood, when I knew that nothing that ever happened in nearby Gatwick Airport would ever destroy the beauty of that place. It's because there are enough people who don't want it to happen. There is, we have to believe, enough love in the world – for the woods and the wild things – to put an end to what once seemed unstoppable and inevitable. We don't have to make false choices between ancient woods and service stations. There is, surely, going to be a perfectly acceptable, alternative brownfield site for Extra MSA's toilets, shops and petrol pumps. Perhaps it's right next to Smithy Wood, in the middle of the half-empty business park. Of course it is.

Smithy Wood needs rescuing. There's no getting away from the fact that it's not in good shape. In one tiny corner of the wood, where the four-wheel drives have been most active, there are three old tyres, several plastic bottles and cans, crisp packets, pipes, a yellow plastic ball, the chewed top of a yellow plastic cone, half a plastic bumper, an exhaust pipe, carrier bags hanging from the branches, many empty beer bottles, sweet packets and the mangled remains of a birch tree that has been torn off at the roots. But that is all the more reason for smothering this place with love, for clearing the undergrowth of the rubbish, for bringing back vitality and living joy to the woods. It can happen almost overnight. Everything it needs is already in the wood and the soil: the seed and fungi, the saplings and the teeming underground fauna. Nature is resilient, if we'll just leave her alone for long enough, and it's easy to see how Smithy Wood can once again become a place of beauty and life. Surely we can plant and preserve enough woods, everywhere, so that people can even drive their 4x4s, or gun each other down with paint, as well as stand and marvel in tranquillity. If we are serious about getting people into the woods, because that's the only way that anyone ever learns to care about them, then we have to make them available for all sorts of activities. We should unleash the off-roaders and the paintballers into their own corners of the conifer plantations. But let's not grub up any more irreplaceable ancient woodland to make way for yet another service station.

We want to believe that anyone spending enough time in the woods will learn to love them. As the twinkly old forester in *The Children of the New Forest* says: 'This is a bad world, and I thank Heaven that I have lived in the woods.' We like to think that the woods cultivate goodness, and that even the most corrupt heart will soften and melt in a sun-dappled glade, and maybe it will, but it's probably best not to take it for granted. We will always need to fight for our woods. People are busy, and others don't care, and where there's money to be made the woods will always suffer. There are many people who hate and fear the wild places. In times of uncertainty, or turmoil, they take their chances. In the early 1660s, as the new parliament argued about how best to preserve Britain's supply of timber, and in particular the incredible reserves of trees in the Forest of Dean, a man called Sir John Winter simply cut them all down. In the year 1661 there were thirty thousand mature oaks in the Forest of Dean; six years later there were fewer than two hundred. This man had lived among the trees all his life – he was a landowner and an ironmaster – but he liked money more.

Our woods are better protected these days, but we don't have enough of them, and we cannot afford to lose any more. The naturalists have been saying this for years. Here's Paul Sterry writing in 2007 in the inspiring but sober setting of the *Collins Complete Guide to British Trees*, all of a sudden sounding like that soldier in *Apocalypse*

Now who snaps and disappears into the jungle: 'Sell the house. Sell the car. Sell the kids…':

> For me, considering how little we have left, the loss or degradation of any more ancient semi-natural woodland is wanton vandalism. To combat further decline, I would urge anybody interested in trees to donate as much money as possible to conservation bodies for the purchase of land. In this context, the only organizations that I would consider donating money to personally are the Wildlife Trusts, the Woodland Trust, the National Trust and the Royal Society for the Protection of Birds.

Indeed. Sell everything! Protect the land! It's not like there isn't a plan to save and conserve and spread our woods. There is – it's very simple and it has been elegantly explained many times by a parade of persuasive and well-informed people. Forests have been felled to make the books that tell us how to save them. At least the tide has turned since the Locust Years, and even the Forestry Commission has now added a hefty portion of native broadleaf trees into its planting mix. The Woodland Trust alone has over 500,000 members. And yet, for all that, the destruction goes on. We lost one hundred ancient woods in the last decade and there are currently over seven hundred more under threat from development, and not just from hubristic railway plans, but from everyday neglect, golf courses, out-of-town centres, and any number of illusory schemes. Over the same period, in just ten years, three-quarters of our butterfly species have declined and tens of millions of birds have disappeared from the woods. We know all this, even if we'd rather not.

The pressure is relentless, but we have to believe that the devastation can be halted and reversed. So here's my own, brief action plan, arrived at during my year in the woods, and based on the conclusions of a host of better-qualified experts. There's a list of groups and people (and where to send your money) at the back of this book. In the end, the solution is simple, even if we all know how hard some of it (the important part) is going to be.

It starts easily enough, though, with a statement of the bleeding obvious…

- Plant hundreds of millions more trees: on the streets and in the woods; on the roadsides, pavements, fields, parks, hills, valleys and lanes. Everywhere. But don't forget that meadows and moors are also priceless.
- Create more woods. Never mind thirty minutes between doughnuts on the motorway, no one should live more than thirty minutes from a wood. We should aim for thirty per cent woodland cover in the UK. And then some.
- Harvest the conifer plantations. Plant them again (adding a healthy mix of broadleaf trees). We need woods of every kind.
- The trees we plant should be local to the area. They are better suited to withstand the coming storms.
- Create nature corridors and stepping stones (bands of woodland, heath and hedgerow; isolated ponds and copses) between the new and existing woods. Bring them right into the centre of our cities. Leave room for nature to breathe.
- Learn to leave well alone. Often, doing nothing is the best way forwards.
- End the large-scale, commercial import of saplings. We can grow our own.
- Fund nature charities so they can buy more woods.
- Or buy your own! Over sixty-five per cent of woods are in private hands in the UK.

- Don't let the government sell off the woods. They're our woods – and they should be held in trust for future generations. Pressure them (and the Forestry Commission) to plant native species and to meet their tree-planting commitments. Right now, they are failing to plant what they have promised.
- Open up the woods to everyone. Fight for access. Keep the footpaths clear.
- Make school trips to the woods part of the National Curriculum. Give every schoolchild in Britain at least one tree to plant. Bring back the timber from a local wood and use it in woodworking classes. Build camps in the forest. Support forest schools. Inspire and engage the next generation.
- Get to the woods! Bring everyone you know: walking, bird-watching, tree-climbing, biking, hiking, camping, swimming, star-gazing, truffle-hunting, and fuel-, herb- and berry-gathering. Live off the land with the woodland survivalists. Ask for a woodland burial. Go paintballing (in a conifer plantation). But get to the woods.
- Fight to save the woods that are left. Every wood is special, but the ancient woods (those we can date back to the year 1600 in England and Wales and the year 1750 in Scotland) are irreplaceable. At present ancient woods take up just two per cent of the UK's land surface. We cannot afford to lose any more.
- Every ancient tree should be placed under a preservation order.
- The idea that 'economic growth' can trump all other considerations when reviewing planning applications is an open invitation to the locusts and the bulldozers… and is based on faulty maths (see below). Reverse the recent legislation.

- Protect and revitalize the green belts. Don't let people tell you they're needed for essential development – or that they're run down so we might as well use them anyway. Our public-spirited forebears had more sense than us.

- There are people who devote their lives to finding ways around planning laws. It is a lucrative business (as one company website puts it) to 'take on planning appeals that can't – or shouldn't – succeed'. Let's close every loophole – and plan an effective deterrent.

- NIMBYs are YIMBYs. We should celebrate and thank anyone who cherishes the place in which they live. Do we sneer at indigenous peoples who fight for their land?

- Even so... the inhabitants of one threatened, leafy village are often the same people whose money, businesses and lifestyles are scorching the fracking zones of Yorkshire and the forests of Brazil. We need to connect to the world beyond our own horizon.

- The natural environment is free in a company balance sheet, but it has a *real* value: Natural Capital. Trees give us life. Are we saying that's worth nothing? If, for now, we have to live in a world run by accountants, then at least let's make sure they get their sums right. This applies to the rain and the rivers, the air and the oceans, the bees, the soil and much more. All of it has value. Make the polluter pay.

- Put woods – and wood – at the heart of everything we do and get our woods working again. More coppicing. More local timber used for industry and homes. More homes in woods! More charcoal. Trees are generous and wood is a flexible, versatile and strong material. Plant more trees and chop them down. Grow the woodland economy – and create jobs.

- Bring back the beaver and the boar. Limit the deer and control the sheep. Let the lynx and the wolf roam free(ish).
- Support the rewilders – those who are trying to bring back the wild into our lives. There's Glen Affric in Scotland, where the Caledonian Forest is being brought back to life; and Knepp Castle, a rewilded farm estate in West Sussex; but anyone with a scrap of land can help.
- Consume less. End our addiction to fossil fuels and growth-at-all-costs. The biggest danger to our woods is man-made climate chaos, fuelled by our hunger for stuff we don't need.
- Protect the woods under threat in any way you can. Boycott the offending companies. Write to your council and MP. Chain yourself to a tree. Consider the possibility that economic sabotage may be the only language that some people understand.
- Call the police if you see someone tipping a broken fridge down a woodland verge. Or, maybe, on the next moonless night, think about putting a rusty medieval mantrap in their way. Bring back the Forest Laws, in all their savagery, but this time for the vandals and the polluters.
- Listen to John Evelyn. The 'improvident wretches' who needlessly tear down our forests, woods and trees should be handed over 'to the vengeance of the Druids'. They can put them in a wicker man and hang them from a forest oak.
- Pray to the gods to punish anyone who desecrates our woods. When Erysichthon chopped down the holy groves, driven by an insatiable lust for timber, and swung his axe against Ceres's sacred oak, the Earth goddess sent *Famine* to him in the night and 'she breathed herself into him, covering his throat, and chest, and lips, with her exhalations', so that when he woke there was nothing that could satisfy his

raging hunger. He ate everything – 'in the midst of eating
he demands to eat' (does this remind you of anything?) – but
still he craved more. He sells all he has, even his daughter,
but nothing is enough for his burning belly. Until, in the
end, Erysichthon 'began to tear at his limbs and gnaw them
with his teeth, and the unhappy man *fed*, little by little, on
his own body'. Or so says Ovid.

- If the gods, or Gaia, won't answer our call, perhaps we
  should take matters into our own hands. According to
  James Frazer in *The Golden Bough*, the ancient Germans had
  a straightforward way of dealing with anyone who attacked
  their sacred groves, one that is worth bearing in mind the
  next time a developer puts in an application to replace an
  ancient wood with an oil rig. For anyone who 'dared to
  peel the bark of a standing tree, the culprit's navel was to
  be cut out and nailed to the part of the tree which he had
  peeled, and he was to be driven round and round the tree
  till all his guts were wound about its trunk... it was a life
  for a life, the life of a man for the life of a tree'.

There. That should do it.

Anyway. An ancient wood for a service station. What kind of
sick exchange is that? The rain is hammering down through the
trees, but it is easy to see that Smithy Wood is ready to rise again, if
only it can be given the chance. Bizarrely, from down the hill in the
business park, and despite the shattering storm, an ice-cream van
has started chiming the tune of 'Food Glorious Food'. I think about
poor starving Oliver and then Erysichthon and his unassuageable
hunger for MORE. We need, urgently, to devise a system of living
that allows people to give full rein to their creativity and zest for
testing every limit, but also encourages us to want to live within

our means. We can't just keep on looting and burning and guzzling everything in sight, all the time hoping that something will turn up – a technological miracle, maybe, or a new planet – that means we never have to say 'enough!' or curb any of our appetites. Is there a plan? I sure do hope someone is working on one...

Most people hate the idea of limits. Especially in the West, where throughout the Cold War it was seen as a patriotic duty to give full rein to our inner toddler, grabbing what we could, shaking our new toys in the faces of those dour, over-regulated drones in the East. And just look at what happens, we were told, if you give too much power to the government and the regulators: you end up with business leaders wrapped in red tape and university lecturers mucking out pigsties in the Chinese countryside. The climate-change deniers of North America are convinced that the world's efforts to reduce their fossil fuel consumption are in reality a United Nations plot to bring the US economy to its knees. Mao and his cultural revolution is their great cautionary tale: it's what happens if you allow the state to interfere in your life, or inhibit your freedom to buy and grow and breed and consume. All of this leads to a simple-enough question: are individual rights and unfettered growth fundamentally at odds with the preservation of the earth's resources? And can we, at this very local level, save Smithy Wood and the other ancient woods of Britain, without imposing on people's freedom? The freedom, for example, of the shareholders of Extra MSA Group to build their service station? Well, put like that, of course we can. We are free to set any limits we want. We just need to be clear about our values.

If you want to know what a life lived without limits looks like, there's always the example of the 'great predator' of the Forest of Dean, Sir John Winter, who burned with enough destructive energy to blight the forest for generations. More recently, though, and on the other side of the coin, there's the self-styled 'publisher, poet and tree-planter', Felix Dennis, who went from being one of the co-defendants at the *Oz* magazine trial in 1971 (when the judge idiotically described him as 'very much less intelligent' than the others, and for that reason less culpable) to becoming one of the world's most successful publishers, responsible for dozens of magazines and websites. He just couldn't help making money, he once explained, but nor did he bother stopping himself from reaching out and grabbing more or less anything else he wanted. He drank to excess, of course, and smoked his cigars, and got himself addicted to crack cocaine; he even somehow survived legionnaire's disease (because, he said, he was seasoned by years of hard living); but he also became a best-selling poet of grace and power and, in 1996, he planted the first tree in something he initially dubbed the 'Forest of Dennis' at the southern edge of the ancient (and vanished) Forest of Arden in Warwickshire. By the year 2013 over one million trees had been planted in Dennis's forest, which now stretched to three thousand acres. His vision (backed up by stupefying quantities of cash) was to create a joined-up area of woodland covering thirty thousand acres and containing over ten million trees. It is well on its way.

Felix Dennis died in 2014 at the age of sixty-seven, but he left his fortune to his wood (renamed the Heart of England Forest in 2011 in a rare and misguided moment of restraint). I've come here straight from Smithy Wood, on this longest day of the year, and I'm standing in late sunshine in the newly opened car park looking at a dozen young saplings in their snug plastic sheaths: by the end of the day I'll have walked past thousands of the things, but even they are still just a tiny fraction of the total frenzy of planting. The woods I walk through are a patchwork of old and new, remnants of copses and plantations joined to fields of new plantings and wild flower meadows. The spirit of Felix Dennis is everywhere. In fact, not just his spirit: on every noticeboard there is an aerial portrait of the man's face, picked out in lush green, his hair a flowing tangle of trees, his beard a mighty forest. He looks like a bespectacled Green Man – the *genius loci* – here to protect his creation. The first of the woods, just the other side of the road from the small car park, is 'Dorothy's Wood', and a winding green sunlit path (surely it should be yellow brick?) curls past hundreds of youthful trees standing proud in a sea of tall thick grass freckled with large daisies. Every single tree is a broadleaf. I see sweet chestnut, beech, maple, rowan, cherry, ash, hazel, hawthorn and oak. There's a little cluster of apple trees and clearings with wild flowers, grasses and commemorative plantings. (A plaque reading 'Gary. Congratulations on your 6th Year. February 2016' is hung over one sapling.) Large truck tyres are sunk into the grass and I think of Smithy Wood, but these ones have been left deliberately for children to clamber over – everything here is fresh and alive with butterflies and the only sound is birdsong. It's like the woods of Narnia after the snows have melted, a young wood in a new world, with dog roses cascading in the hedgerows and not a conifer in sight. There's an ease and a joy to starting things afresh in the sunshine. Poor old battered Smithy Wood.

Around another corner and there's another baby forest, this one only about five years old, with the young trees having only just shed their protective containers – they're stacked under a mature oak like so many discarded snakeskins. The trees are now about double the height of the summer grass. There's whitebeam here, as well as oak; the cherry is growing the fastest, I see, as well as some of the ash trees. It's such a joyful sight. *Excelsior!* Onwards and upwards. The path leads through an older wood, although the planters have been busy here too. There's a vast willow collapsed over a stream, its exposed root base the height of two tall men. Sunlight is sifting through the pale root ends and a lump of thick Warwickshire soil flops into the water with a sudden splash. Stratford-upon-Avon is not far from here, and although the Saxons probably felled most of the original Forest of Arden, it's a warming thought that Shakespeare may have ridden or walked this way. Arden – or Eden by any other name – is being pieced back together, sheltered by a wealthy man's dream, far from the roiling crowds.

> Sweet are the uses of adversity;
> Which, like the toad, ugly and venomous,
> Wears yet a precious jewel in his head;
> And this our life, exempt from public haunt,
> Finds tongues in trees, books in the running brooks,
> Sermons in stones, and good in everything.
> I would not change it.

> SHAKESPEARE, FROM
> ACT II, SCENE I, *AS YOU LIKE IT*

These woods have been designed for joy. There are dozens of streams (or is it just the one, coiling through the trees, flitting under the footbridges?). There are benches everywhere, in small clearings, backed up against ancient trees, and planted at the top of every sweeping view. In a reversal of the great slaughter of the 1870s, the agricultural land is being returned to forest. The thought pops into my head that my father, always alert to fruitful industry, would be unutterably happy wandering among these woods. The close-packed path emerges into another meadow, sheltering a large pond. Dragonflies hover among the sun-streaked reeds and I sit down gratefully on a new, bleached-wood bench. Thank you, Mr Felix Dennis and 'David William Graham, beloved Husband,

Father and Grandfather', whose bench this is. You have conjured a place of peace.

And I wonder: is that what this is? A peaceful haven, carved out by a man who could afford to keep the world at bay, if only for a while. And is this the best chance our woods have to survive and flourish for generations to come? The random involvement of people with money to spare? Felix Dennis was predictably relaxed about his own efforts here: 'Will the trees make a difference? No! Nature doesn't care. She'll shrug once we're gone.' But if you think for a moment about who (or what) makes the best guardian of the land, it's hard not to be nervous about most of the options. To rake it all up again: Britain has so much less woodland than almost any other European country and our wildlife is in catastrophic decline. We've just about halted the large-scale felling of woods (and the large-scale planting of conifers), and about twelve per cent of Britain is now woodland (although Dennis sneered at the unrealistic optimism of this figure), but we still have a huge amount to do. Do we just wait and hope for a thousand more Forests of Dennis?

I wouldn't trust the government. They can be blown in almost any direction by short-term panic or a need for funds. In 2011 the Coalition government's environment secretary suggested selling off their share of the UK's forests for £250 million. In times of austerity, Caroline Spelman suggested, it was the government's duty to raise a bit of cash – that's about as much money as it costs to build twelve kilometres of dual carriageway – and if the woods are then developed into dirt-track racing venues, well that's economics. There was a huge outcry, Spelman apologized, and the government backed down. We love our woods, but we can't afford to lower our guard. All over the world, from the tar sands of Canada to the forests of China, hungry governments (democratic and otherwise) are consuming their own limbs. Monarchs and dictators have been

no better: they may have reserved some of our loveliest forests
for their own pleasure, but they've felled just as many when their
coffers needed swelling. The same, unfortunately, has always been
true of the landowners – some of them may have wanted to tend
their estates for the future benefit of their heirs, but that heir was
just as likely to tear down the lot to fund his high living. Countless
local people must have watched in agony as some feckless lord sold
off yet another wood – and deprived them of their fuel and the
food for their pigs.

> Shall I have jealous thoughts to nurse,
> When I behold a rich man's house?
> Not though his windows, thick as stars,
> Number the days in every year;
> I, with one window for each month,
> Am rich in four or five to spare.
> But when I count his shrubberies,
> His fountains there, and clumps of trees,
> Over the palings of his park
> I leap with my primeval blood;
> Down wild ravines to Ocean's rocks,
> Clean through the heart of No-man's Wood.
>
> W.H. DAVIES, 'NO-MAN'S WOOD'

The enclosure of the land, and the mass migration of people from
the country to the cities, robbed the woods of their most ardent
defenders. More often than not, the people who relied on local
woodland for their fuel and food were the ones most likely to rally
to its defence when the landowner wanted to grub it up to pay his
debts. The 'Tragedy of the Commons' is not the one laid out in
1968 by Garrett Hardin in his hugely influential article of the same

name – the idea that if we all share a wood or a field of wheat, I'll get up early one morning and take as much as I can for myself. That petty bit of classroom theory has been thoroughly debunked by Oliver Rackham, among others; it's not what happened in the real world. Free from outside disruption, people were careful to share their resources, and were perfectly capable of nurturing their grazing lands and woods so that everyone had their fair share. If someone seemed to be taking too much, or despoiling the common land, they were restrained by the majority (through a sense of responsibility, or the law, or sometimes, presumably, by a poke in the eye with a sharp stick). The real tragedy is that most of us no longer have enough direct contact with the woods (and the farms, fields, meadows, commons and moors) and we have no idea that they are so neglected, or in so much peril. To know them is to love them.

Should people be able to buy and sell woods? Sometimes the idea seems absurd. But since the days of the Saxon kings, for over one thousand years, every inch of Britain has had an owner. Perhaps we need a 'No-man's Wood' – a vast fairy-tale forest that doesn't belong to anyone. Perhaps this is what the government should be doing with its (with *our*) woods. In fact, isn't this, in effect, what we have already? National forests, held in trust by a succession of governments, for all eternity. Their only duty is to tend them, harvest the timber and keep them open to the public. As for the rest of the UK's woods, Sian Atkinson of the Woodland Trust believes we have a good mix – government, businesses, individuals, charities – and the important thing is to nail down the legal protection. In fact, the real truth, the truth we need to take to our hearts, is that the best guardians of the woods are every single one of us: wood owners, ramblers, citizens and voters, National Trust members, tree-planters, tree-huggers, grannies, children, bird-watchers, anglers, runners…

I'm snapped out of this reverie of list making by an elderly couple, who wander over to my bench from the pond where they've been throwing stale bread at the fish. The sun is drifting to earth on this day of magic, the longest day of the year. Their dog scuffles after them through the tangled grass. 'Isn't this such a perfect place to sit,' says the woman, 'if you have things you need to think through? And too many troubles in your life.' Was I looking that agitated? Or muttering to myself? I ask if they knew Felix Dennis. 'Oh yes,' she says, 'we met him a couple of times. Such a nice man. We live in a village not far from here, and some of the footpaths are only open to "locals", so we asked at the estate office and Mr Dennis rang us that very night to say we'd be welcome any time. He didn't need to do that. Did you know he spent over £22,000 on fireworks for the village every year? And have you seen his memorial? And his mother's? She died soon after him, you know. No? Well, I'll say no more, you need to see it, but it's not far, just through the woods. We'd often bump into him, walking here, looking at his trees.' The man eases his way in. 'We need to plant more trees,' he says, 'we can all agree on that, I hope.'

A path of clover leads gently uphill through the fields towards Felix Dennis's memorial. On the other side of a low hedge there are high mounds of neatly rolled hay, and behind them, and in every direction, there are more saplings, nursed in their plastic rolls, every single one of them looking sappy and green with vigorous young life. I'm so glad I came to this wood. And there, at the top of a low hill, set in a wide strip of grass, surrounded by banks of young trees, is what looks like a large sandstone menhir, and next to it, with his back turned, is a more-than-lifesize, green-and-brown resin statue of Felix Dennis, shoulders hunched, surveying his forest. A couple of his poems are here, carved in bronze on the menhir and engraved on a clear glass plaque. 'I plucked all the cherries', one starts. Too right you did, I think, although apparently we're 'welcome to them now'. From the front, the Statue of Dennis has one hand in his trouser pocket and the other is holding a small pot, containing a tiny oak tree. After all his millions, and his roller-coaster life, all he really cared about was his forest. The words of the man I met by the pond have followed me here: 'We need to plant more trees.'

Or, as the resident poet once wrote:

> Whosoever plants a tree
> Winks at immortality.

We can all agree on that, I hope.

Epilogue

If you should lose me
Oh, yeah, you'll lose a good thing.
You know I love you,
Do anything for you.
Just don't mistreat me,
And I'll be good to you.
'Cause if you should lose me,
Oh, yeah, you'll lose a good thing.

I'm giving you one more chance,
For you to do right.
If you'll only straighten up,
We'll have a good life.
'Cause if you should lose me,
Oh, yeah, you'll lose a good thing.

This is my last time,
Not asking any more.
If you don't do right,
I'm gonna march out of that door.
And if you don't believe me,
Just try it, daddy,
And you'll lose a good thing.
Just try it, daddy,
And you'll lose a good thing.

BARBARA LYNN,
'YOU'LL LOSE A GOOD THING'

Touch Wood

Croft Ambrey, late May, twelve months later

It is a year since I was last here in Croft Ambrey, and the world is already a hotter, drier, wetter and more combustible place. It's lonelier, too. In the past twelve months, while I've been wandering in the woods, we've lost somewhere between one thousand and ten thousand species worldwide. (It's so hard to be precise about all those beetles and bugs, but in the days before humanity started consuming everything in sight, including its own limbs, the rate of loss, the 'normal' background rate, was about five – yes, *five* – species per year.) People who follow these things, the naturalists, conservationists, biologists and environmentalists, never seem to know whether to shout and scream, or feed us little morsels of good news in an attempt to get us to change our ways. To be frank, neither approach seems to be working that well, at least not at a global level. If the earth really is going to come to her own defence, and brush us off like a bad case of dandruff, as James Lovelock once suggested in *Gaia*, she's certainly leaving it late.

The National Trust sign is still here, letting us know that the conifers' days are numbered, although at some point someone has moved it further down the hill and into a fresh area of operation. There doesn't seem to have been much new clearance work done. Perhaps they're waiting for the scars to heal, and the new broadleaf saplings to start growing, before they rev up the chainsaws once more. An ancient, full-grown sweet chestnut tree – one of those that were apparently planted to commemorate the defeat of the Spanish Armada – looms over the dividing fence, a battered reminder of what these young saplings may one day grow into. It has been butted and bruised by cattle and is covered in knobs, whorls, burls and cracked branches and it looks more like an Ent, one of Tolkien's talking, walking trees, than anything I've ever seen. There's a rabbit hole at its base and a bird's nest cradled above my head. It's a friendly tree.

My children, who are not young, are admiring the pinewood. It's just like Narnia, they tell me, the trees standing tall against this blue, cloudless sky, and it makes them think of Christmas days and wild, Russian forests. But listen, I try to say, there's nothing growing on the dead forest floor, and these conifers are foreign aliens, planted in soul-sapping industrialized rows, but they look at me with pity and alarm when I mention foreigners. Is our father a xenophobe? There's no denying that the pinewoods do smell bracing and fresh.

The churned and macerated dead pine that last year covered half the hillside has mulched and softened and is blending with the landscape. Tentative brushes of grass have spread from the edges and – deeper in – there are new outposts of bracken, buttercups, thistles, nettles, dock leaves, red campion, dandelions, nodding stands of tall mauve foxgloves and even some pioneering bluebells on the verges. There's a lark shrieking overhead. I peer into one of the fenced areas reserved for the broadleaf saplings and count two oaks, a rowan, a hazel and a tiny hornbeam, all of them tenuously alive in their cosy plastic sheaths. Across the rest of the reservations, about half of the new trees seem to have survived the year – and that's surely enough, with rain and sun and luck, to bring back the broadleaf forest. I wonder what would happen if every human on the planet were to fall asleep for one hundred years like the princess and her courtiers in *Sleeping Beauty*. The mass extinctions would end. The forests would return. And here on Croft Ambrey we'd be left with a jumble of conifers and young broadleaf trees, scrambling towards the sky, bothered by deer and bewildered flocks of sheep. Will they miss us when we're gone? And who would tell them how beautiful they are?

It's a crystal-clear day at the top of the Iron Age fort and there must be a record-breaking number of counties in view. There's a busy rumble from the quarry, but away from that, and in every direction as far as I can see, there is rich farmland, villages, distant purple hills and, most noticeably, woods: strips and squares of conifer plantations and ragged green splashes of broadleaf forest. Skylarks and buzzards dance and drift in the sky. The hawthorn is blossoming and the ash is on the brink of leafing, late as ever, but latent with pent-up force. Something ancient, known and named by Wordsworth, settles in the air:

> ... And I have felt
> A presence that disturbs me with the joy
> Of elevated thoughts; a sense sublime
> Of something far more deeply interfused,
> Whose dwelling is the light of setting suns,
> And the round ocean and the living air,
> And the blue sky, and in the mind of man;
> A motion and a spirit, that impels
> All thinking things, all objects of all thought,
> And rolls through all things.

FROM 'LINES WRITTEN A FEW MILES
ABOVE TINTERN ABBEY'

Ah, Gaia... are you there? Everything looks perfect and untroubled on this blue, cloudless day and I am alone in 'a silent place that once rang loud', as Edward Thomas wrote of another lonely hillside. If we look out over the fields, and into the distant woods, we won't see what is no longer there: the butterflies and birds, the lynx and the boar, the orchids, the fungi and, of course, the people. The woods (or what is left of them) are emptier than they've been for millennia.

Regardless, the vast old oak is still here, on the edge of the summit. Of course it is. I have walked down the hill a short distance, when I am hit with an overpowering urge to reach its trunk and to feel its embrace (whatever it may say), but it's not easy from the path below: the ground is rutted and thick with brambles and I'll have to circle round and get to it from above. My time is short, though, and I'm expected somewhere else ('the world is too much with us') and I hesitate on the path before heading up and round the tree, trying to find a way to its heavy centre. Clouds of insects are hanging in its shade and deep swathes of nettle and bramble have closed off every approach. There is no path, so I just have to pull my way through, stumbling over thorns, pricked and stung by nettles, briefly wedging my foot in a hidden rabbit hole. But I'm here, now, at the heart of the tree, under the shelter of its massive branches, and I can see that it's a hollow tree, with two great trunks forking at head height. I won't touch, I think, I'll bow to this tree that exudes such power, and has stood here for so many centuries, and whose great-grandfather was once worshipped, perhaps, by the people who lived all those centuries ago at this Iron Age fort.

There is so much we still have to learn about trees, but one thing we can say is that they do 'talk'. Not to us, so far as we know, but to each other. They emit chemical signals to warn each other of predators – everything from aphids to elephants – so that they can produce terpenes to attract wasps to eat the aphids, or tannins to make their leaves distasteful to large mammals. When a tree is attacked, every other tree downwind will react to the threat. But there is so much more that we don't know, despite our centuries of cohabitation. I wonder, in fact, if we are now more ignorant about the hidden lives of the trees than our Iron Age ancestors ever were. Most of us don't seem to realize – or we seem to have forgotten – how interwoven trees are with our own survival.

It is so quiet here and there is no human sound in the air. It feels wild and strange, under this steepling oak, bathing in its strength only a couple of paces from its dark centre. I have a sudden urge to hug the thick, grooved trunk. Close up, it is dauntingly huge. I step forwards and rest my cheek on its rough skin and then lean my arms and chest against its immense body. It is so soothing, like nothing I have ever experienced. A surge of energy races through my limbs. I can feel a restless agitation being drawn from me and then into the tree and up and away. More than that. It's as though some toxin or poison is being drained and nullified. There is an incredible connection of intensity and power – and it probably only lasts for a second or two before, shockingly, the tree sneezes, violently. Or, if it's not a sneeze, it's a sharp, explosive sound of distress or anger or an exhalation of pain. I spring back and think, well, that's just a squirrel protecting its nest or (ludicrously) maybe it's an adder. I try again, and it happens again – a violent shriek of protest. I know, really, that it's a squirrel making that noise, it must be, but I also know that I've had my brief moment with this tree and it's time to pay my respects and disappear. We've both had our say – and I feel free of something twisted and dark that I hadn't even known was there. It troubles me though, as I walk down the hill feeling lighter than I have for years, after the most fleeting of connections with this ancient tree, it troubles me that the debt is all on my side.

Thanks to...

Rebecca Winfield, for the help and support, and without whom this book would not have been written; the brilliant Sam Carter, of Oneworld Publications; Adam and Gill Beaumont, for the poetry book; my aunt Biddy, for words of sceptical encouragement; my brother Toby, for The Map and much else; Ruth Segal; the eagle-eyed Amanda Dackombe; Jacqueline Hutson, for an eye-opening walk around Markstakes Common; Sian Atkinson of the Woodland Trust; Oliver Newham, also of the Woodland Trust; Emma and Nick Vester; Andy Tunmer; the inspirational Judy Ling Wong; Roz Marston and her wood-loving friend, Annie Bailey; Paul Wood; Jonathan Bentley-Smith; Matthew Frith of the London Wildlife Trust; Barbara Lynn and Frank Lipsius; and of course, above all others, to Anna, Natalie, Alex and Esme, for all their suggestions, and for putting up with a distracted tree-obsessive in their midst. Especially Anna, who had the best advice of all: 'just stop research-ing, and start writing!'

Acknowledgements

I have quoted from many writers and poets in these pages and am extremely grateful to all the copyright holders who allowed me to use their words. I made strenuous efforts to secure permissions (reluctantly jettisoning some favourites) and apologize if there are any omissions or mistakes. Please get in touch if so! You'll find most of the sources in the Bibliography, but I would also like to acknowledge the following:

'Teddy Bears Picnic', song, Jimmy Kennedy (1932).

'The Woodland Thicket Overtops Me' (Anglo-Saxon poem), quoted in Della Hooke, *Trees in Anglo-Saxon England* (The Boydell Press, 2010); Kenneth Jackson, 'Writing in the Wood', *Studies in Early Celtic Nature Poetry* (1935).

'Afforestation' by R.S. Thomas, from Rupert Hart Davis, *The Bread of Truth* (Grafton Books, 1964).

Selections from H.L. Edlin appear in *Trees, Woods and Man* (Collins, 1956) and *British Woodland Trees* (BT Batsford Ltd, 1944).

T.H. White, *The Sword in the Stone* (GP Putnam, 1939).

Eva Salzman, 'Ending up in Kent', *Double Crossing: New & Selected Poems* (Bloodaxe Books, 2004).

Line from 'Woods' by Louis MacNeice, from *Holes in the Sky* (1948); see Louis MacNeice, *Selected Poems* (Faber & Faber, 2007).

Prose selections from Edward Thomas appear in *One Green Field* (Penguin Books, 2009) and *The Woodland Life* (Forgotten Books, 1897).

'Walnut St., Oak St., Sycamore St., Etc' Copyright © 1980 by Wendell Berry, from 'Ronsard's Lament for the Cutting of the Forest of Gastine Woods'. Reprinted by permission of Counterpoint.

The Lorax by Dr Seuss reprinted by permission of HarperCollins Publishers Ltd © Dr Seuss (1957).

The Lion, the Witch and the Wardrobe by C.S. Lewis © copyright CS Lewis Pte Ltd 1950.

Jim Robbins, *The Man Who Plants Trees* (Profile Books, 2013); first published in the United States in 2012 by Spiegel & Grau, an imprint of Random House, Inc; also published in the UK as *The Man who Planted Trees: A Story of Lost Groves, The Science of Trees, and a Plan to Save the Planet* (Spiegel and Grau, 2015).

Gerald Wilkinson, *Trees in the Wild* (Book Club Associates, 1976).

Jacqueline Memory Paterson, *Tree Wisdom: The Definitive Guidebook to the Myth, Folklore and Healing Power of Trees* (Thorsons, an imprint of Harper Collins, 1996).

Robert Lancelyn Green, *The Adventures of Robin Hood* (Puffin Books, 2016). Text copyright © Robert Lancelyn Green, 1956.

Alfred Noyes, 'A Song of Sherwood', reprinted by permission of The Society of Authors as the Literary Representative of the Estate of Alfred Noyes.

Lucy Goodison, *Holy Trees and Other Ecological Surprises* (Just Press, 2010).

Melusine Draco, *Traditional Witchcraft for the Woods and Forests* (Moon Books, 2012).

Yvonne Aburrow, *The Enchanted Forest: The Magical Lore of Trees* (Capall Bann Publishing, 1993).

Danu Forest, *Celtic Tree Magic: Ogham Lore and Druid Mysteries* (Llewellyn Publications, 2014).

Poem by Dafydd ap Gwilym from *A Celtic Miscellany*, translated by Kenneth Hurlstone Jackson (Penguin Books, 1971).

Simon Schama, *Landscape and Memory* (Harper Collins, 1995).

Ron Freethy, *Woodlands of Britain* (Bell & Hyman, 1986).

Collins Complete Guide to British Trees, reprinted by permission of HarperCollins Publishers Ltd © Paul Sterry (2007).

Felix Dennis, 'Whosoever Plants A Tree', taken from *Tales From The Woods* (Ebury, 2010), reprinted by kind permission of the Felix Dennis Literary Estate.

Lyrics of 'You'll Lose a Good Thing', by Barbara Lynn Ozen, used by permission of Jamie Music Publishing Co., Philadelphia, PA, USA.

Select Bibliography

I have bought too many books about woods and trees over the years. These are the ones that were most useful, and inspirational, when I was researching and writing my own.

BOOKS ABOUT WOODS AND TREES

Baker, Richard St. Barbe, *I Planted Trees* (Lutterworth Press, 1944)

Campbell-Culver, Maggie, *A Passion for Trees: The Legacy of John Evelyn* (Eden Project Books, 2006)

Collis, John Stewart, *The Wood* (Penguin Books, 2009)

Deakin, Roger, *Wildwood: A Journey Through Trees* (Penguin Books, 2008)

Edlin, H.L., *British Woodland Trees* (B.T. Batsford Ltd, 1944)

Edlin, H.L., *Trees, Woods and Man* (Collins, 1956)

Evelyn, John, *Sylva; Or a Discourse of Forest Trees, and the Propagation of Timber in His Majesties Dominions* (Classic Reprint, 2012)

Freethy, Ron, *Woodlands of Britain: A Naturalist's Guide* (Bell & Hyman, 1986)

Gill, Charlotte, *Eating Dirt: Deep Forests, Big Timber and Life with the Tree-Planting Tribe* (Greystone Books Ltd, 2011)

Gilpin, William, *Remarks on Forest Scenery and other Woodland Views* (first published in three volumes 1791, available from Forgotten Books)

Irving, Henry, *How to Know the Trees* (Cassell & Co, 1910)

Mabey, Richard, *The Ash and The Beech* (Vintage, 2013)

Marren, Peter, *The Wild Woods: A Regional Guide to Britain's Ancient Woodland* (The Nature Conservancy Council, 1992)

Miles, Archie, *Hidden Trees of Britain* (Ebury Press, 2007)

Milner, J. Edward, *The Tree Book: The Indispensable Guide to Tree Facts, Crafts and Lore* (Collins & Brown, 1992)

Pakenham, Thomas, *The Company of Trees* (Weidenfeld & Nicolson, 2013)

Pakenham, Thomas, *Meetings with Remarkable Trees* (Weidenfeld & Nicolson, 1996)

Penn, Robert, *The Man Who Made Things Out of Trees* (Penguin Books, 2016); see also his BBC4 series *Tales from the Wild Wood*

Rackham, Oliver, *The Ash Tree* (Little Toller Books, 2015)

Rackham, Oliver, *Woodlands* (HarperCollins 2010; originally published in 2006 as *Woodlands*, volume 100 in the New Naturalist series)

Robbins, Jim, *The Man Who Plants Trees* (Profile Books, 2013; published in the United States in 2012 by Spiegel & Grau, an imprint of Random House, Inc.)

Scott, T.H. and Stokoe, W.J., *Wild Flowers of the Wayside and Woodland* (Frederick Warne & Co., Ltd, 1936)

Starr, Chris, *Woodland Management: A Practical Guide* (The Crowood Press, 2013)

Sterry, Paul, *Collins Complete Guide to British Trees* (HarperCollins, 2007)

Thompson, Andy, *Native British Trees* (Wooden Books Ltd, 1998)

Wilkinson, Gerald, *Trees in the Wild and other Trees and Shrubs* (Book Club Associates, 1976)

HISTORY AND SCIENCE

Cannadine, David, *G.M. Trevelyan: A Life in History* (Penguin Books, 1997)

Cartmill, Matt, *A View to a Death in the Morning: Hunting and Nature Through History* (Harvard University Press, 1993)

Culpeper, Nicholas, *The Complete Herbal*, first published in 1652 as *The English Physitian*.

Hayman, Richard, *Trees: Woodlands and Western Civilization* (Palgrave Macmillan, 2003)

Hooke, Della, *Trees in Anglo-Saxon England: Literature, Lore and Landscape* (The Boydell Press, 2010)

Hoskins, W.G., *The Making of the English Landscape* (Penguin Books, 1970)

Paxman, Jeremy, *The English: A Portrait of a People* (Michael Joseph, 1998)

Perlin, John, *A Forest Journey: The Role of Wood in the Development of Civilization* (Harvard University Press, 1991)

Rackham, Oliver, *The Illustrated History of the Countryside* (George Weidenfeld and Nicolson Ltd, 1994)

Schama, Simon, *Landscape and Memory* (HarperCollins, 1995)

Stringer, Chris, *Homo Britannicus: The Incredible Story of Human Life in Britain* (Penguin Books, 2006)

Tudge, Colin, *The Secret Life of Trees: How They Live and Why They Matter* (Penguin Books, 2006)

Wood, Michael, *In Search of England: Journeys into the English Past* (Penguin Books, 2000)

POETRY AND LITERATURE

I have quoted from the great woodland writers John Clare, William Wordsworth, S.T. Coleridge, John Keats, Tennyson and others – but their poetry is, as they say, widely available from all good book shelves.

Cobbett, William, *Rural Rides* (Penguin Classics, 1985)

Drabble, Margaret, *A Writer's Britain* (Thames & Hudson, 2009)

Fiennes, Celia, *Through England on a Side-Saddle* (Penguin Books, 2009)

Fowles, John, *The Tree* (Vintage, 2000)

Graves, Richard Perceval, *A.E. Housman: The Scholar Poet* (Oxford University Press, 1981)

Hardy, Thomas, *The Woodlanders* (Oxford World's Classics, 2009)

Hurlstone Jackson, Kenneth, *A Celtic Miscellany: Translations from the Celtic Literatures* (Penguin Books, 1971)

King, Angela and Clifford, Susan (eds) *Trees Be Company: An Anthology of Poetry* (Green Books, 2001)

Knox, Collie (ed.) *For Ever England: An Anthology* (Cassell & Co, 1943)

Sheers, Owen (ed.) *A Poet's Guide to Britain* (Penguin Classics, 2009)

Somerville, E. OE. and Ross, Martin, *The Irish R.M.* (Abacus, 1990)

Stowell, Leonard (ed.) *The Call of the Open: A Little Anthology of Contemporary and Other Verse* (A&C Black, 1922)

Taplin, Kim, *Tongues in Trees: Studies in Literature and Ecology* (Green Books, 1989)

Thomas, Edward, *Collected Poems* (Faber & Faber, 2004)

Thomas, Edward, *The Woodland Life* (first published 1897, available from Forgotten Books)

Thomas, Edward, *The Heart of England* (1906) and *The South Country* (1906), selected and published as *One Green Field* (Penguin Books, 2009)

Thomas, R.S., *Collected Poems 1945–1990* (Weidenfeld & Nicolson, 2000)

Young, Andrew, *A Prospect of Flowers: A Book about Wild Flowers* (Penguin Books, 1986)

ENVIRONMENTAL CONCERNS

Avery, Mark, *Inglorious: Conflict in the Uplands* (Bloomsbury Natural History, 2015)

Griffiths, Jay, *Wild: An Elemental Journey* (Penguin Books, 2008)

Klein, Naomi, *This Changes Everything* (Penguin Books, 2015)

Monbiot, George, *Feral: Rewilding the Land, Sea and Human Life* (Penguin Books, 2014)

MYTHS, FOLKLORE AND MAGIC

Aburrow, Yvonne, *The Enchanted Forest: The Magical Lore of Trees* (Capall Bann Publishing, 1993)

Aldhouse-Green, Miranda, *The Celtic Myths: A Guide to the Ancient Gods and Legends* (Thames & Hudson 2015)

Briggs, Katherine M., *British Folk Tales and Legends: A Sampler* (Granada Publishing, 1977)

Draco, Mélusine, *Traditional Witchcraft for the Woods and Forests* (Moon Books, 2012)

Forest, Danu, *Celtic Tree Magic: Ogham Lore and Druid Mysteries* (Llewellyn Publications, 2014)

Frazer, Sir James George, *The Golden Bough: A Study in Magic and Religion* (Oxford World's Classics, 2009)

Goodison, Lucy, *Holy Trees and Other Ecological Surprises* (Just Press, 2010)

Graves, Robert, *The White Goddess* (Faber & Faber, 1999)

Hughes, Ted, *Tales from Ovid* (Faber and Faber Limited, 1997)

Maitland, Sara, *Gossip from the Forest: The Tangled Roots of Our Forests and Fairytales* (Granta Publications, 2012)

Memory Paterson, Jacqueline, *Tree Wisdom* (Thorsons, an imprint of HarperCollins, 1996)

Philpot, Mrs J.H., *The Sacred Tree or The Tree, Religion and Myth* (Macmillan & Co, 1897; available from Forgotten Books)

Pullman, Philip, *Grimm Tales: For Young and Old* (Penguin Classics, 2012)

Westwood, Jennifer and Simpson, Jacqueline, *The Lore of the Land: A Guide to England's Legends, from Spring-Heeled Jack to the Witches of Warboys* (Penguin Books, 2005)

CHILDHOOD

Briggs, Barbara, *Our Friendly Trees* (The Lutterworth Press, 1933)

Kipling, Rudyard, *Puck of Pook's Hill* (Macmillan & Co, 1914)

Lancelyn Green, Roger, *The Adventures of Robin Hood* (Puffin Classics, 1995)

Lewis, C.S., *The Lion, The Witch and The Wardrobe* (HarperCollins Children's Books, 1998)

Marryat, Captain, *The Children of the New Forest* (originally published 1847)

Seuss, Dr., *The Lorax* (HarperCollins, 1997)

White, T.H., *The Sword in the Stone* (GP Putnam's Sons, 1939)

Resources

If you want to know more about woodland conservation – or have ever dreamed of owning a wood – here is a list of organizations to get you started. They're in no particular order, except that I did want to put the Woodland Trust first:

Woodland Trust

www.woodlandtrust.org.uk

They've planted over thirty-six million trees and saved more than five hundred woods since they were formed in 1972, but there is so much more still to do. Support them! And visit their woods (easily found on the website).

National Trust

www.nationaltrust.org.uk

As well as coastal paths, old buildings and estates, the National Trust conserves many hectares of ancient woodland.

National Trust for Scotland

www.nts.org.uk/Home

More of the same. In Scotland.

Small Woods Association
www.smallwoods.org.uk

If you own a small wood, or would like to, and want to know how to make it flourish, then this should be your first port of call.

· Trees for Cities
www.treesforcities.org

They plant trees in cities. Heaven knows, we need them.

The Wildlife Trusts
www.wildlifetrusts.org

A network of forty-seven Wildlife Trusts, run by local communities and set up to work for nature's recovery on land and at sea. They are always looking for volunteers.

Black Environment Network (BEN)
www.ben-network.org.uk

Working for greater ethnic environmental participation. There's a great list of publications and resources to be downloaded for free.

RSPB
www.rspb.org.uk

They buy and nurture woods (and wetlands, meadows and coastal sanctuaries) so that birds and humans can flourish.

Woodlands.co.uk
www.woodlands.co.uk

Woodlands for sale!

National Forest
www.nationalforest.org

Over two hundred square miles in the centre of England: once derelict land, old open-cast coal mines and neglected farms, now planted with over eight million trees, as well as meadows and lakes.

Heart of England Forest
www.heartofenglandforest.com

Felix Dennis's legacy, a glorious new broadleaf forest growing in the heart of England (*see* Chapter 10). But it's no longer called the 'Forest of Dennis', alas.

Trees for Life
www.treesforlife.org.uk

Founder Alan Watson Featherstone's mission is to restore the great Caledonian Forest – helped by hundreds of volunteers every year. Over one million trees planted, mostly Scots pine, juniper, aspen and birch. The wild animals should follow, we hope.

Rewilding Britain
www.rewildingbritain.org.uk

Bring back the beaver! And the lynx, elk, wild boar and wolf. We need more wild spaces in our lives.

Knepp Castle Estate
www.knepp.co.uk

An experiment in rewilding and 'wild-range' farming on a 3,500-acre estate in Sussex.

Forestry Commission
www.forestry.gov.uk

Once, they butchered and poisoned the native forests and blanketed our land with conifer plantations. Nowadays they are taking a more measured approach, and considering aesthetics and leisure amenities as well as fruitful forestry. It's good to keep reminding them...

Botanical Society of the British Isles (BSBI)
www.bsbi.org

For all your wild flower needs. Inspirational.

Ancient Tree Forum
www.ancienttreeforum.co.uk

They seek out, cherish and protect the UK's ancient and veteran trees.

Sherwood Forest Trust
www.sherwoodforest.org.uk

Protecting, conserving, sustaining and growing the forest of the future.

ConFor
www.confor.org.uk

Short for the Confederation of Forest Industries – in other words any organization or individual who is concerned with the business of forestry.

Continuous Cover Forestry Group
www.ccfg.org.uk

CCFG promotes visually and biologically diverse woodlands and is 'committed to advancing "close to nature" sivicultural systems'.

Natural England
www.gov.uk/government/organisations/natural-england

The government's adviser for the natural environment in England, 'sponsored' – it says on their website – by the Department of Environment, Food and Rural Affairs (DEFRA).

Ramblers Association
www.ramblers.org.uk

Champions of walking and walkers, keeping our footpaths clear.

CPRE
www.cpre.org.uk

The Campaign to Protect Rural England. Home of the YIMBY – just say yes to green belts, dark skies and beautiful English countryside…

Save Our Woods
www.saveourwoods.co.uk

Formed in response to the UK government's proposed sell-off of public forests. It is leading a very successful campaign, but the fight goes on.

Sylva Foundation
www.sylva.org.uk

An environmental charity working to revive Britain's wood culture.

Tree Council
www.treecouncil.org.uk

Grew out of the successful 'Plant a Tree in '73' scheme – a charity that encourages us to plant more trees. But there's more to it than that now.

Woodland Heritage
www.woodlandheritage.org

Founded by a group of traditional cabinet makers, who wanted to encourage the proper management of woods and trees. They fund research and conservation projects.

The Native British Trees

A native British tree is most often defined as one that managed to colonize mainland Britain after the end of the last Ice Age (about ten thousand years ago) and before the waters rose cutting off these islands from the continental mainland (probably about three thousand years later). So you would think, wouldn't you, that by now people would have had enough time to agree on a definitive list, but, predictably, it's not that simple. Not only are the early years lost in a murk of confusion (did the beech manage to claw its way on to these shores at the last minute – or were its nuts paddled across many years later by Stone Age hunters?), but there has also been much recent rearrangement of the species and subspecies and an agonized debate over what constitutes a hybrid. And there's even a question mark hanging over what we mean by 'native'. Perhaps we should include anything that made its way here without being handled by a human? For all we know, a bird may have brought the seed of the first spindle, many years after the English Channel was formed. And what, come to think of it, is a *tree* anyway? There are plenty of people who would include the purging buckthorn in our list of trees; others place it firmly as a *large shrub*. Dogwood, too – and… a difficult one, this… what about the elder?

Anyway, here's a list of large (they can reach at least five metres), perennial, woody plants with a branching crown (supported by a primary stem), which are most likely to have arrived in mainland Britain before it became an island. There are twenty-six of them, although you may well wonder why I have separated the aspen

(*Populus tremula*) from the black poplar (*Populus nigra*) and what all those sorbuses are doing on their own; or indeed what the elder is doing here at all, when the spindle (*Euonymus europaeus*) is left languishing as a shrub. There are five evergreens (the box, holly, juniper, the Scots pine and the yew) and the rest are broadleaves. Not a larch or a sitka spruce in sight…

The fact is I *wanted* there to be twenty-six native British trees. That was the number I was told long ago, and it seems pleasing that there are the same number of British trees as there are letters in the alphabet. This also ties in neatly with what we know of the Early Irish language, ogham, whose twenty (later twenty-five) letters were once believed to have been ascribed to (or inspired by) different species of tree or shrub. You can read an especially baroque celebration of this theory in Robert Graves's 'The White Goddess', and in particular in his reimagining of the medieval Welsh poem 'The Battle of the Trees', although he later claimed he'd become unhinged through overwork and dismissed the whole thing.

The field guide I use is the brilliant *Collins Complete Guide to British Trees* by Paul Sterry.

- alder (*Alnus glutinosa*)
- ash (*Fraxinus excelsior*)
- aspen (*Populus tremula*)
- birch: silver (*Betula pendula*); downy (*Betula pubescens*)
- beech (*Fagus sylvatica*)
- black poplar (*Populus nigra*)
- blackthorn (*Prunus spinosa*)
- box (*Buxus sempervirens*)
- cherry: wild (*Prunus avium*); bird (*Prunus padus*)
- crab apple (*Malus sylvestris*)
- elder (*Sambucus nigra*)

- elm: smooth-leaved elm (*Ulmus minor*); wych elm (*Ulmus glabra*)
- hawthorn: common (*Crataegus monogyna*); midland (*Crataegus laevigata*)
- hazel (*Corylus avellana*)
- hornbeam (*Carpinus betulus*)
- holly (*Ilex aquifolium*)
- juniper (*Juniperus communis*)
- lime: small-leaved (*Tilia cordata*); large-leaved (*Tilia platyphyllos*)
- maple (*Acer campestre*)
- oak: pedunculate or English (*Quercus robur*); sessile (*Quercus petraea*)
- rowan (*Sorbus aucuparia*)
- Scots pine (*Pinus sylvestris*)
- whitebeam (*Sorbus aria*)
- wild service tree (*Sorbus torminalis*)
- willow: bay (*Salix pentandra*); crack (*Salix fragilis*); white (*Salix alba*); almond-leaved (*Salix triandra*)
- yew (*Taxus baccata*)

Index